LESS SUCKS

Overpopulation, Eugenics, and Degrowth

Peter Coffin

This book was adapted from the script of a Very Important Documentaries film of same name. To watch Less Sucks: Overpopulation, Eugenics, and Degrowth, visit:

YOUTUBE.COM/@IMPORTANTDS

"Overpopulation" (like the simultaneous "overproduction") is only relative to the capitalist conditions of production. [This] reactionary and vicious propaganda [conceals], under [the] cover of obsolete clerical superstitions, the true social causes of poverty and misery.

-RAJANI PALME DUTT, FASCISM AND SOCIAL REVOLUTION (1934)

INTRODUCTION

Are we, humanity, doing our duty to save the planet from ourselves? Should we live in the middle of nowhere, in the tiniest of tiny houses? Does our individual consumption put out demands that the market will ultimately supply?

That's why everyone should check their "species privilege" by going vegan, ceasing contributing to the 0.025% of plastic in the ocean made up of plastic straws by using paper ones, and going number two outside, producing fertile soil for the garden where, over the course of a year, we can produce enough food to feed a single human being for a couple of weeks. All tomatoes.

Of course, people will all have a few guilty pleasures. Like when we run out of tomatoes almost immediately, someone must go into town and buy groceries. While in town, why not catch a movie from the Marvel Cinematic Universe?

One of the great MCU films is *Disney's The Muppet Christmas Carol* by Charles Dickens, the story of a man named Ebenezer Scrooge. In the beginning, Scrooge is a greedy banker, which is anti-semitism. To counter that, let's acknowledge the noble, virtuous profession of banking.

Anyways, by the end, Scrooge was giving away turkeys like that question mark suit guy claims the government gives away grant money.

But anyways, one day, a pair of Robert Owen-types come in and ask Scrooge to donate some money for their efforts to feed the poor. That beautiful tightwad said no and that the poor should go to workhouses, and workhouses SUCKED. Basically, a poor person could trade labor for a place to stay, but the living conditions were intentionally grim so only "truly" dirt poor people would do it.

These do-gooders replied with (and this is a direct quote that exemplifies how well Dickins holds up today), "what the hell, bro? Most people would rather die than do that shit."

Scrooge shot back, "if they would rather die, they had better do it and decrease the surplus population, bruh."

Ebenezer Scrooge knew how important it was to degrow the economy because there just aren't enough resources to support this many people. It's not necessarily because there are too many people, mind you, but because there are not enough resources!

Charles Dickens' *Disney's The Muppet Christmas Carol* came out back in 1849 – the early days of the MCU when it was just Iron Man. And even back then, the world was overrun with people and we've got literally 7.5 times as many people now!

How did the world ever get so... overpopulated – so full!? There's just nowhere to go!

...

Enough with the act. I think overpopulation and degrowth are anti-human BS, I think they're tied to eugenics, and I don't think Less is More. Less is less, and all people deserve more. Further, anti-growth ideology is meant to cover up the major flaw inherent in capitalism, one that causes crisis after crisis.

Less Sucks, so cultivate some mass, we're about to take a huge dump on Malthusianism.

PART 1: MALTHUS AND OVERPOPULATION

"Surplus population" is a term Charles Dickens has his famed tightwad Ebenezer Scrooge use to establish that the character agreed with the then-popular ideas of a man named Thomas Robert Malthus. The term refers to "unproductive consumers," a segment of the population providing nothing "socially necessary" while also using up valuable resources. Welfare queens, Netflix employees, and people who don't seed torrents for at least 24 hours after they're done downloading. Those people should just die, I guess!

Anyways, Malthus, an unfortunately real person that Dickens didn't make up for a story, published *An Essay on the Principle of Population* in the year 1798.

Before we go into that, though, it's important to know that Malthus wasn't the *first* person to raise a version of this concern. Centuries earlier, Plato, noted writer of Socrates fanfiction and democracy-hater, had expressed concerns about overpopulation when earth only had around 150 million people on it (less than 2% of today's population).

In one of his most prominent works of Socrates fan fiction, *Republic*, Plato put forward that population growth was unsustainable – not only because of the demand for food and resources, but he believed it was one of the biggest causes of war.

The thing that should have clued people in on how full of shit Plato is was his idea that in a just world, philosophers would rule because "only *they* understand what is good – and what is 'good' can not be clearly seen or explained."

Naturally, Plato had the idea that the philosopher-kings would impose legal restrictions on reproduction – let's call them "Bang Rules" – to create balance and harmony. Obviously, the philosopher-kings weren't subject to the Bang Rules, though, and were allowed to bang without restriction. I mean, we want more eternal smarties, right? And fewer idiots buying tubs of pre-made chocolate frosting to eat as a snack, right? How could that go wrong?

---

458 THE REPUBLIC, V. 271

"Clearly then we shall next endeavor to make marriages as sacred as possible, and those most advantageous to the State will be sacred."

"Unquestionably."

"How then will they be most advantageous? Answer me that question, Glaucon. For I see in your house hunting-dogs, and a great number of game birds. Pray now, tell me, have you given any attention to their mating and breeding?"

"In what particular way?" he asked.

"Why, in the first place, among these creatures, although all are of fine stock, are there not some which are, or prove themselves to be, the best?"

"There are."

"Well, then, do you breed from them all indiscriminately, or do you take special care to breed from the best?"

"From the best."

"And do you pick out the youngest, or the oldest, or only those in their prime?"

"I select only those in their prime."

"And without such precautions in pairing do you think the breed of birds and dogs would deteriorate greatly?"

"Yes, I do."

"And do you think the case is at all different with horses and other animals?"

"It would be absurd to believe it," he replied.

"Good heavens, my dear friend," I said, "how preeminently skilful must our rulers be if the same principle holds with regard to the human race!"

Plato wanted to achieve a "stationary state" in terms of population, believing it would end hunger, class conflict, and even wars of conquest. This would be good for Plato (a philosopher) because then no one would want to oust the philosophers, who would rule over everyone *because they knew so darn much.*

By the time Malthus's work became popular in the early 1800s, the human population had grown to around one billion people, well over six times what Plato thought could be sustainable. Making Plato look like a bitch.

Still, Thomas Robert Malthus reintroduced to popular concern the idea that humankind would soon outgrow available resources. Like Plato, Malthus proclaimed that a finite amount of land would be incapable of supporting a population with supposedly infinite growth potential. Which eventually made Malthus… also look like a bitch.

So let's just quote this bitch verbatim. In the first edition of 1798's *An Essay on The Principle of Population*, Malthus asserted that:

> *Population, when unchecked, increases in a geometrical ratio. Subsistence increases only in an arithmetical ratio. A slight acquaintance with numbers will shew the immensity of the first power in comparison of the second.*
>
> – Robert Thomas Malthus, An Essay on The Principle of Population (1798)

To break it down, Malthus's original claim is that while increases in food and resource production occur in a linear fashion, increases in population occur exponentially.

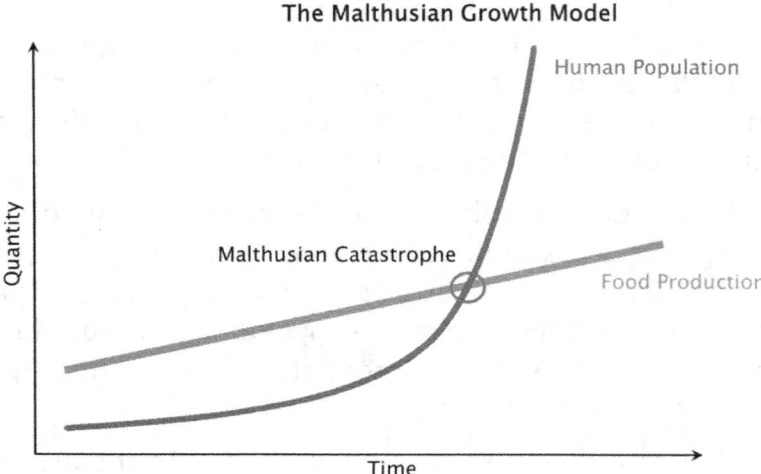

Which makes sense, because industry only ever scales in a linear fash-

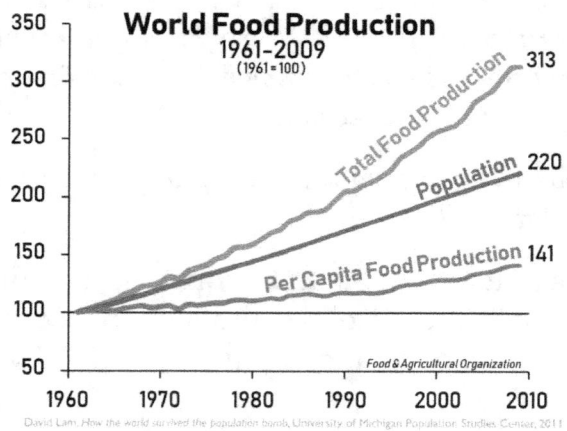

Okay whatever, but the human population endlessly expands exponentially forev–

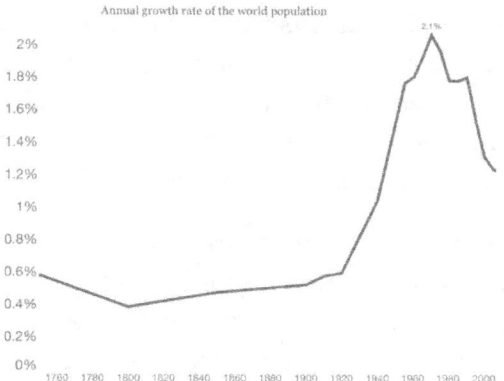

Many, from OG Dialectical Materialist Karl Marx to Julian Simon, a pro-market liberal environmental economist of the Chicago tradition (one of the most disciplined and concise critics of overpopulation the *Limits to Growth* study resurrected Malthusianism in the 1970s[1]), have been able to recognize that graph as... flimsy.

In *Theories of Surplus Value*, written in 1863, Karl Marx extensively debunks Malthus's body of work. He not only noted that Malthus was a plagiarist, but also that he misrepresented (and even sometimes outright misunderstood) the theories he plagiarized, then changed up his justifications whenever called out. Marx wrote:

> "[Malthus's] theory was taken from [James] Anderson. His Principles of Political Economy was a polemic work written in the interests of the capitalists against the workers and in the interests of the aristocracy, Church, tax-eaters, toadies, etc., against the capitalists. Where [Malthus] inserts his own inventions, it is pitiable."

> "Malthus's [work] was a lampoon (mockery) directed against the French Revolution and the contemporary ideas of reform in England (Godwin, etc.). It was an

> *apologia for the poverty of the working classes."*
>
> *"Furthermore, Malthus [took] the Andersonian theory of rent to give his population law [...] an economic and a real (natural-historical) basis, while the nonsense about geometrical and arithmetical progression [...] was a purely imaginary hypothesis."*
>
> *"Anderson had defended premiums on [corn] exports [...] and duties on corn imports not out of any interest for the landlords, but because he believed [it] "would reduce the average price of corn" and ensure even development of the productive forces in agriculture.*
>
> *"Malthus adopted [...] Anderson's [assertion because] he was a professional sycophant of the landed aristocracy, [to justify their position] economically. Malthus defends the interests of the industrial bourgeoisie only [when they're] identical with the interests of [...] the aristocracy, i.e., against the mass of the people, the proletariat. But where these interests diverge and are antagonistic to each other, he sides with the aristocracy against the bourgeoisie. Hence his [assertion] of the "unproductive worker", overconsumption etc."*
>
> *– Karl Marx, Theories of Surplus Value (1863)*

So here's where we begin to understand what Malthus was ultimately doing: he's not defending capitalism, he is slyly attacking it – not for the right reasons, mind you. Rather than identifying and criticizing the contradictions of the system and ultimately siding with the working class, he was telling industrial capitalists to KNOW THEIR PLACE in defense of the property-owning aristocracy, the church, and ultimately the feudal mode, which was crumbling and needed some support, you guys (uwu)!

When the interests of the working class are in direct

conflict with *both* the industrial capitalists and the aristocracy, Malthus is happy to defend the capitalists. However, the capitalists are ultimately beneath the aristocracy and, in his eyes, their true purpose should be to serve the aristocracy.

Today, we live in a world where the capitalists won. The feudal aristocracy has been replaced by the bourgeoise, the previous middle (and current ruling) class. While the class for whom Malthus was a "professional sycophant" for has all but symbolically disappeared, his theories have persisted.

They haven't become any less wrong, though.

We produce an excess of food nowadays, so much that a lot of it goes to waste. One-third of food produced for human consumption is lost or wasted globally. This amounts to about 1.3 billion tons per year, worth approximately US$1 trillion.[2]

Because of this, arguments similar to Malthus's on food are often put forward with the substitution of generalized resources for food production (which, like I said, is a severely outdated concern), though still contrasted with the world human population.

In *Capital Volume 3*, Marx talked of the implementation of "momentary and forcible solutions of the existing contradictions, violent eruptions, which restore the disturbed equilibrium for a while." Basically, the present state of affairs becomes untenable as conflicts and discrepancies build, then a series of events pacify these contradictions, albeit at a great human cost.

In response to crises, the capitalist ruling class implements policies that range from propping up massive failures of institutionalized finance capital at the expense of the worker (both here in America and abroad) in the wake of the 2008 crisis to, well, both world wars. Prosperity at the cost of human life in one way or another; anarchically violent and destructive means are utilized to temporarily restore stability without actually ending any of the fundamental contradictions that generate these crises.

The boom part of the cycle was obviously mirrored in the human population. I mean we called it the Baby Boom! It's the sole event in human history that truly embodies the conditions that the idea of a Malthusian catastrophe puts forward as necessary for one to happen. But, much to the chagrin of hunter-gatherer fetishists, society didn't collapse!

As productive forces developed, the standard of life increased and the family model responded. People started having two kids rather than 6, partly because both parents began working, and partly because the kids were surviving; feeding six of them is a lot of work. Population regulated itself; birth rate tapers off and becomes linear after the 60s, despite growth in the production of food (and other resources) outpacing human population growth.

When there are more people, linear growth (having the same number of births as the previous year despite the existence of more people of child-bearing age than before) means the birthrate actually went down.

More people having fewer kids.

We don't call that the Baby Bust, though, do we? That might remind people of that boom-and-bust cycle that pervades society as long as capitalism's fundamental contradictions persist!

But, thing is, people were noticing them anyways. The 1960s were rife with crises of imperial-stage capitalism, from groups oppressed in the imperial core demanding civil rights to the imperial state waging an unpopular war in a newly formed communist state.

The New Deal and the World War that powered it only averted a crisis and temporarily made things look shiny and new. But the rot beneath, the inequalities and the perceived scarcity that created them caused a reaction...

David Lloyd George, Prime Minister of Britain during the First World War, came to a peculiar realization as the Great

Depression kicked off:

> "If we had not had a great war, if we had gone on as we were going, I am sure that sooner or later we would have been confronted with something approximately like the present chaos. There must be something fundamentally wrong with our economic system, because abundance produces scarcity."
>
> - David Lloyd George, former UK Prime Minister, April 7, 1933

David Lloyd George identified that there was a flaw, but not what it was. The fundamental contradiction of capitalism is that everything made (as well as all of the wealth generated from it) is retained by the class that owns the means to produce it. This is the heart of all inequality; production is where power comes from in modern capitalist society because it is through production that all things happen. What Lloyd George and everyone else missed was just the capitalist ruling class doing what a capitalist ruling class does.

So really, this "scarcity" is perceived by the underclass because *it is exerted on them*. Further, it reinforces an ideological line: that everything is limited and "the good life" is only for a select few – those who "prove themselves" in the Great Free Market.

So, the wealth of this vast world is accumulated and concentrated among a tiny class of capitalists. This prevents the workers, their subjects, from not only simply accessing the earth's resources, but even understanding how they work.

In his 2021 book, *Breadtube Serves Imperialism*, activist and journalist Caleb Maupin addresses these more generalized "limited resources" arguments as such:

> *What is the inherent flaw regarding the logic of "limited resources"? This logic fails to take into account that the way human beings interact with*

> *resources has constantly been in a state of change. Across Africa, minerals are being extracted and sold at high value currently, when just several decades ago they were worthless. Computer chip technology has rendered them very valuable. The long predicted "peak oil" scenario in which petroleum would run out causing a global catastrophe never panned out, because new methods of extracting oil, from deep sea drilling to hydraulic fracking have been invented.*
>
> - Caleb Maupin, Breadtube Serves Imperialism (2021)

Probably the most valuable aspect of what Maupin says here is the lack of value judgment; he simply asserts the way society uses resources is always in flux. It's neither good nor bad; it's just true.

The takeaway is that there is no stationary "law" to our relation to natural resources. As our technology develops, we *should* have more choice as to how resources are used – and which ones to use.

The true issue is that everything, from the media disseminating information to the mining equipment pulling resources out of the ground, to the factories where production occurs, is owned by a class of people with a set of material interests that contradicts everyone else's. So this is all done as cheaply as possible and sold back to us at a premium; not only are they ripping us off, but they're polluting the earth.

If we don't like that? Tough. That's the capitalist mode of production.

If We The People had the power, wealth, and ownership, we would HAVE to make decisions "democratically" – there would be no alternative. Would we be ripping ourselves off? Polluting the world without actively converting to a consistent, zero-carbon baseload (like nuclear) and plentiful, renewable alternatives for the various other fuels and energy we need?

Would we would vote to swim in our own shit? I don't think so. However, we do not have the opportunity to vote "no" on that. Sure, we can certainly say "that's how I would do it," but we don't have any power. The owning class does.

This is all the result of the way relations of production have arranged society. There is no fundamental truth that necessitates this, no decree from God, nothing more than an unspoken social contract that was signed long before any of us were born... so why don't we just change it?

Well, Malthus' principle of population is easy to adopt as a framework to claim the way society has historically played out isn't the result of structures that can be changed as criticism unearths their fundamental contradictions, but rather the result of supposedly unquestionable laws of nature.

In fact, there's a good chance this is a significant reason why many people think "capitalism works because of human nature."

This paints the capitalist world as a "natural order" where the "superior" humans – *the owners and directors of production* – will naturally rise to the top and the inferior ones – *the unproductive consumers* – will sink to the bottom. The inequalities inherent in capitalism and previous socioeconomic systems are therefore supposedly explained, and all those doing "nothing of merit" need to "know their place."

It's not hard to see how the people at the top would benefit from this view of how things are. These types statistically pollute the most, as evidenced by even slippery charts trying to diffuse blame among the top 50% (see OXFAM CO_2 charts) by talking about "lifestyle consumption emissions."

Nice try, Oxfam, it's still visible that the top 50% do barely anything compared to the top 1%! And even so, that's not addressing where the power really comes from here!

But seeing that the capitalist ruling class funds

propaganda to explain why the earth is so bad, it makes perfect sense they would look at the issues of sustainability and longevity of the human species and see population itself as the problem, rather than flying a private jet everywhere, demanding bluefin tuna decorated in edible gold garnished with lobster while their fracking investments pay off because their other investments in the supply chain paid off and the interests of one feed into the other and so on and so on.

Why face the idea it might be inefficient modes of production and distribution – ones that might waste and pollute more but are cheaper and yield higher margins – when they benefit more from just... not facing it.

In 2009, some of these owning class folks, David Rockefeller, Bill Gates, George Soros, Warren Buffett, and others, formed something they, for some batshit insane reason, decided to call "The Good Club." As in, "if you go against our club, you're clearly Bad, because we are Good. Obviously! It says so in our name!"

The global health portion of their focus was on disease and... overpopulation. The Guardian, a left-leaning news outlet most consider to be mainstream, characterized pushback on this as "conspiracy theories" that are "coming from the American Christian Right." That's right, to be concerned about anything Malthusian-sounding was characterized by the mainstream left in 2009 as a "religious, right-wing conspiracy"[3].

Which is bad! This Club, as clearly stated before, is Good! They're very forward-thinking and progressive. They just want the best for themse... all of us! And that means knowing who to cull!

It's very convenient for the capitalist class (or media outlets they own) to characterize any pushback on their assertion that "there are too many people" as "anti-progressive." If the progressives think that the problem is population, then progress, the thing progressives want, seems like it might just be addressing the surplus population.

Capitalists would never be considered "the surplus population" in *capitalism*. Again, they're *supposedly* the cream that's risen to the top. They *supposedly* deserve everything they have. They're *supposedly* more vital as people, leading us towards a better future with their capital. They invest in the technologies of tomorrow! They aren't disposable. They shouldn't be the ones who starve if there are "too many people" for the capitalist mode of production to sustain!

So, a lot of progressives blame people who are *not* capitalists. Instead, the problem is "people who have children and buy food at Dollar Tree." Whether they realize it or not, they are defending the mode of production that generates inequality and isolates the decision-making process of all of these things to the capitalist ruling class.

Keeping in mind that perspective, it's very easy to see why socioeconomic systems rooted in these ideas would be kept in place. It's "privilege" and insurance all at once. It's comfort, it's power. There's no incentive to end these systems, and these viewpoints will never truly die under them. Capitalism is the supposedly logical answer to questions of humanity's essence – "human nature," the supposedly immutable, unchanging natural law at the center of all of us.

But "human nature" changes alongside ideology[4], which develops dependent on the mode of production as it creates contradictions that need justifications. What we call "human nature" in capitalism is an assumption informed by the fact most people will not have enough, and therefore can never get enough. The false scarcity the capitalist ruling class must create to compensate for their oversaturated markets depends on their subjects accepting that assumption.

But everyone could have enough. More than enough. And we'd have different assumptions if we knew, for sure, that we would never go hungry.

When people stop believing in these ideologies – either due to apparent contradiction, changing conditions, or systemic crisis, the capitalists really start noticing how big that "surplus population" is.

So, when Malthus asserted endless, exponential population growth was inevitable, the solution he put forward was to regulate the population. He gave two ways this could be done: either the death rate goes up or the birth rate goes down.

When Scrooge said to let the poor die off to "decrease the surplus population," it was a reflection of the popular interpretation of Malthus's work. The poor have always been blamed for the conditions they lived in and when discussing this in relation to overpopulation, it was said that helping the poor would worsen conditions for everyone in the long run. From *An Essay on the Principle of Population:*

> *Instead of recommending cleanliness to the poor, we should encourage contrary habits. In our towns, we should make the streets narrower, crowd more people into the houses, and court the return of the plague.*
>
> *In the country we should build our villages near stagnant pools, and particularly encourage settlements in all marshy and unwholesome situations. But above all, we should reprobate specific remedies for ravaging diseases: and those benevolent, but much mistaken men, who have thought they were doing a service to mankind by projecting schemes for the total extirpation of particular disorders.*
>
> *If by these and similar means the annual mortality were increased... we might probably every one of us marry at the age of puberty and yet few be absolutely starved.*
>
> *– Robert Thomas Malthus, An Essay on The Princple of*

Population (1798)

"We don't help those sickening poors! Sure, it *feels* like the right thing to do, but they overpopulate the world if you help them! The poor should live worse and shorter lives to increase the death rate so 'we' can live in abundance!"

"We," of course, being the ruling class Malthus made a living kissing the asses of. Also, is it me or did Malthus throw in a little hint of "actually, it's ephebophilia" at the end there? Was he writing this shit to defend Lord Epstein?

In capitalism, the ruling class is supposedly not ruling because of their lineage – and the poor supposedly have the "freedom" to become rich. An illusion of upward mobility must exist; people have to believe they just need to prove that they're "worthy." Of course, it is a rigged game so most can't, and if they can't, well, they were "part of the surplus population" and looked down upon by the Scrooges of the world as a burden.

If this is starting to sound like eugenics, that's because... it pretty much is!

So yeah, Malthus really could have been writing for Epstein, who wanted not only to fuck kids but also to create a master race with his own "superior DNA" by impregnating women at his ranch. According to a 2019 exposé by The Irish Times, "at one session at Harvard, [Jeffrey] Epstein criticized efforts to reduce starvation and provide healthcare to the poor because doing so increased the risk of overpopulation."

That's why next, we're going to talk about eugenics.

PART 2: EUGENICS

The span in time Malthus, Dickens, and Marx occupied was one of growing consciousness in the underclasses of the capitalist world. In particular, the horrific institution of slavery was increasingly questioned in a country where freedom was a founding virtue. And in that question, the question of "what is freedom" leads many to uncomfortable answers; if, indeed, the United States was a land of opportunity and upward mobility, why were so many poor? Didn't they have the freedom to succeed?

In efforts to preserve slavery and to otherwise keep labor as inexpensive as possible, this all had to be rationalized. This necessitated effective, seemingly-unquestionable ideological suggestions that certain people were superior to others. To accomplish this, several hundred years of philosophy and ideology had to be synthesized.

Plato and his *Republic* carry a lot of generalized responsibility here (and we'll circle back to that shortly). Still, the specific idea of "waste people" being put to "good use" goes back much further than Malthus to the 16th-century British colonization of what would come to be called America.

In 16th-century England, a policy of enclosure, the annexation (theft) of "common land" from "commoners," was formalized in Parliament. These acts of enclosure had been going on since the introduction of the feudal mode, however, and had driven many of the peasants from the countryside into the cities of England with nothing. As they crowded the city streets, open disdain for the poor grew along with their numbers.

Formalized inequality was created by privatizing the

common land. Later, informal inequality was created by appropriating the profits to the owners of industry for them to accumulate. As capital became the dominant power, the feudal mode died a slow death.

At the same time, the primary contradiction of capitalism was making its presence known.

Many had nothing while few had everything. The many that had nothing built up, desperate for work, but without enough work for them to do. Malthus would come to call this the "surplus population," but London-born clergyman William Harrison presented rationalization long before in 1577's *Description of England.*

According to Harrison, rather than wanting to work, these poor were simply idle "by nature." Their wasted energy wasn't a result of an economic condition, it was a failure of the poor to settle and find work. None of this was the result of fundamental contradictions in shifting power relations; they were just "waste people" whose lives amounted to nothing of value.

In Nancy Isenberg's *White Trash, the 400-Year Untold History of Class in America,* the author chronicles (among other things) British colonial policies dedicated to "taking out the trash" or relocating the growing "surplus population" to put these "waste people" to "good use."

This trans-Atlantic human population shift, combined with an aristocratic "fixation on animal husbandry," propagated a specific way of describing the poor "not only [as] waste, but as inferior animal stocks, too."

Ebeneezer Scrooge of *Disney's Marvel's The Muppet Christmas Carol* by Charles Dickens was against helping the poor in any way. He offered the idea that if the poor wanted to die, they should do so and reduce the surplus population. But why would they want to die? Most want to live, maybe even improve their situation over the course of their life.

The reason the poor might "want to die" is because the alternative was staying in workhouses, where they could trade labor for a place to stay. The inhumane living conditions of these workhouses were intentionally grim to make them a choice based purely on survival and no upward mobility of any kind was offered.

But when people start calling each other the "surplus population," implying that there is an "overage of people," it's often an indicator that there's not enough work for everyone. If only there were some kind of "New World" to exploit...

One of the principal promoters of American exploration and exploitation was Richard Hakluyt, who engineered a plan to send this swath of "waste people" of "inferior stock" to the soon-to-be colonies, creating a new, continent-sized workhouse. In terms of the stated goals, this plan was a success. As Isenberg put it in *White Trash*, "the land and poor were cultivated together."

Britain's new workhouse was an easy way to "take out the trash," ultimately converting the trash to a valuable asset by having it cultivate, produce, and ship useful commodities back to Britain. So maybe instead of "taking out the trash," we should call it recycling! How environmentally friendly! Converting waste into something useful!? Were the English the original tree huggers, or what?

Speaking of labor, founding father and 3rd President of the United States, Thomas Jefferson, owned 600 slaves over the course of his lifetime. In his book, *Notes on the State of Virginia*, he tried to rationalize why he was allowed to do that in the country that began when he, himself, wrote the words "all men are created equal" in the Declaration of Independence on July 4th 1776, attributing human equality as a Law of Nature.

That must have been a wild contradiction to carry while wandering around his 5,000 acre Monticello plantation ordering people he owned as property because of the color of their skin around.

> *I advance it therefore, as a suspicion only, that blacks... are inferior to the whites in the endowments of body and mind.*
>
> *- Thomas Jefferson, Notes on the State of Virginia (1785)*

I dunno, if I were kidnapped forced into slavery in, like Finland or somewhere where they don't speak English, I would probably seem uneducated and uncooperative. I'd be really pissed off, too. Black people's perceived disposition under US slavery seems pretty self-explanatory to me. It's not exactly science!

But saying "I suspect black people are less intelligent"... also isn't science. It's a notion. From a guy. A pretty powerful guy, but a guy.

One of the selling points of capitalism was that some guy wouldn't be making all the decisions anymore; the market would. Everything, including ideas, competed in the market and only the best succeeded. For slavery to be justified, there would have to be something that proved it made sense in a country where everyone was born equal. The idea of "black people being inferior" would have to prove its merit, right? Like there would have to be something seen as scientific proof that held up to the scrutiny of the day which made it completely okay to treat black people like work animals.

Sooooooo... craniology.

Around the same time Thomas Jefferson was "suspicious that black people are dumb," a man named Franz Joseph Gall was developing a new field called phrenology. Gall asserted that the brain was made up of 27 individual organs, and that by feeling a person's head, one could reveal their natural tendencies. However, phrenologists didn't view these as limitations. In fact, phrenology developed into a kind of "realize your true potential" self-help thing that can only last so long, and it was considered debunked as a pseudoscience by the 1840s.

Fortunately, there was another, totally different thing called craniology (also known as craniometry) that picked up right where phrenology left off! Phrenology (old and busted) was about the brain having a bunch of organs that dictated personality traits, and one could tell which of the organs were bigger by the shape of a person's skull. But craniology (the new hotness) was about measuring brain size by the skull, and also that the shape of different skulls proved "races" were actually different species.

Craniology was the brainchild of a man named Samuel George Morton, who wanted to disprove that God created humanity as one whole, instead believing that God actually created different species of humans.

Kind of like if someone hated penne, but loved ziti. Yeah, sure, I guess they're different things and one can be used to justify racism... but wow are they categorically the same thing. No one is going to prove that God did either of those things any easier than getting most people to care that Mom used ziti because grocery day is coming up and there's no penne in the pantry.

Also, ziti is racist.

What "tipped Morton off" to this revelation of polygenesis was that the sons of Noah (of ark fame) couldn't possibly account for every race on earth. Obviously, the only logical alternative to that is that all races have been separate species from the very start! Right!?

Now, again, we're just talking about some guy's notions, as Morton was keenly aware of. So, he engaged in an incredibly ambitious project to provide *evidence* for this claim: he collected hundreds of human skulls from all over the world, cataloged them, and measured them. I don't know what he did with the actual skulls after that because A&E's *Hoarders* show didn't exist for another 150-ish years and that is the only way I would be interested enough to look into it.

Whether he let them pile them up in his house or he had several storage units around town to maintain an image of normalcy, I don't know. But he did claim that this was evidence that proved one could judge the intellectual capacity of a race by the average size of the skulls he had measured.

Large skulls, according to Morton, meant a large brain and, therefore, high intellectual capacity. Also, he asserted a small skull indicated the opposite because, like with most body parts, bigger is better and that's that – just ask porn! Also, also, different skull shapes could dictate various personality types. Also, also, also, this is all obviously insane and wrong. No rigorous research would support any of these conclusions.

Both elephants and whales are examples of animals with brains significantly larger than ours, and though they are both very smart and cool (don't want to piss off the elephant or whale fandoms), they can't really compete with humans in terms of intelligence. That's okay, humans can't compete with them in terms of dongs.

So, brain size doesn't actually mean anything in regard to intelligence. Similarly, we have about the same brain-to-body mass ratio as mice, yet mice haven't invented a self-driving electric vehicle that runs into parked cop cars yet. So they aren't smart! With some very basic comparisons, we find that it's neither size nor density that makes a brain exceptional.

Essentially, Morton was full of shit.

Nevertheless, in 1839, Morton published *Crania Americana*, in which he claimed that white people had the biggest (and therefore best) brains while black people had the smallest brains. Science, he said, had proven black folks are "not gud brane peepul and to no hire for smaRt job," to use a direct quote.

The book was also written entirely in crayon.

In 1857, George Gliddon and slaveowner Josiah Nott (followers of Morton's) used their scientific reputations to defend the institution of slavery. They published *Types of Mankind* and made claims saying things like black people achieve their "greatest perfection, physical and moral, and also greatest longevity, in a state of slavery."

Two years later, in 1859, Charles Darwin's work on evolution provided an explanation for how life went from nothing to something in a way that didn't involve God. Now, Darwin himself was kind of wishy-washy on the topic of race, believing races were different enough to categorize on a fundamental level. However, he also held a hard stance that slavery was appalling. But we aren't really talking about Darwin here. Also, not believing in slavery doesn't make cheap or free labor cease to be more profitable than expensive labor, though, does it? I mean, Morton showed that SCIENCE distinguished human races, right? The skulls man, the skulls!

Darwin's theory shifted the conversation away from God, but it was one of Darwin's biggest fans, Herbert Spencer, and Darwin's own half-cousin, Francis Galton, who would apply his concepts of evolution to perpetuate Morton's contention that science could categorize human races by intelligence even as the conversation departed from religion and into more general natural sciences.

In 1864, Herbert Spencer published *Principles of Biology* because it was apparently really *in* to name long-form literature promoting one's notions *Principles of Something*. Now, Spencer was an unapologetic supporter of laissez-faire capitalism as he believed that struggle caused self-improvement, which was also something Darwin's work influenced him to believe is be genetically passed on.

Spencer's idea of "survival of the fittest" (yes, he's the guy who came up with that) was directly influenced by not only Darwin, but (yep) *Principle of Population* – Malthus's work. In fact, on page one, the first sentence one of Spencer's *Theory on Population* is him quoting a guy *saying another guy's* assertion that

Malthus's theory is false *is false*. Now, if that confuses the reader don't feel bad. It sure confused me. But what he's saying, right off, is that Malthus's theory is true.

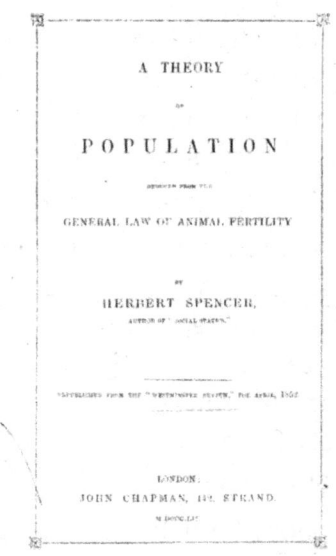

Further, he believed that overpopulation would lead specifically to the fittest taking their rightful place on top, and everyone else would just die out. Basically, Spencer pulled the age-old culture war move of taking whatever negative thing is said about one's beliefs and replying, "yes, but it's good!"

To sum up: Spencer's view was, "there are too many people on earth, and that is fuckin great! The superior white people will know how to use the situation, and the other races will be too stupid to do anything but die off. Hell yeah!"

Herbert Spencer's writings were seen as imperative contributions to "scientific racism," which is the belief that empirical evidence exists to support or justify a belief in racial inferiority or racial superiority. A lot of belief going on there! If that doesn't sound a whole lot like science, that's because it's not. The assumptions of racism have been widely and routinely debunked through the years, especially since the discovery of

DNA. Physically, people of different races just aren't different on any fundamental level, and even so-called "European" skin color has been shown to have genetic origins in Africa[5].

Regardless, the "Spencerian" take on evolution transforms Darwin's work into vindication for market capitalism while ultimately claiming yet another variation of the idea that "savages" are "less evolved" and "incapable of higher thought." Spencer also cited "craniological" studies of skull size and capacity in the tradition of Sam Morton to support this, because of course.

Another thing worth noting happened in the midst of all this. The United States Civil War made slavery in the country illegal... mostly. This didn't stop race science, though. First, because freed slaves were starting with literally nothing and there needed to be ideological explanations for why black people, when free, seemed confined to the lower rungs of the working class in a society where all are supposedly free to succeed.

But secondly, even a lot of people with anti-slavery views didn't necessarily believe all humans to be created equal. For instance, here's Charles Darwin saying that "civilized," aka "western," aka white nations were filled with a race of people with larger brains, larger morals, AND larger dicks than the lower races, which were everywhere else.

> With civilized nations, as far as an advanced standard of morality, and an increased number of fairly well-endowed men are concerned, natural selection apparently effects but little; though the fundamental social instincts were originally thus gained. But I have already said enough, while treating of the lower races, on the causes which lead to the advance of morality, namely, the approbation of our fellow-men—the strengthening of our sympathies by habit—example and imitation—reason—experience and even self-interest—instruction during youth, and religious feelings.
>
> [16] 'Hereditary Genius,' 1870, p. 347.
> [17] E. Ray Lankester, 'Comparative Longevity,' 1870, p. 115. The table of the intemperate is from Neison's 'Vital Statistics.' In regard to profligacy, see Dr. Farr, "Influence of Marriage on Mortality," 'Nat Assoc. for the Promotion of Social Science,' 1858.

On the very same page of *The Decent of Man*, the book this quote comes from, Darwin cites *Hereditary Genius,* the work of English polymath and all-around unpleasant-looking man Francis Galton. Galton took influence from both Darwin and Spencer's ideas, ultimately proposing that modern societies afforded those with "hereditary defects" the ability to survive despite an inability to adapt.

Galton disagreed with Spencer, however, that free markets were a good mechanism for clearing out the chaff, though, because markets developed the productive forces – and well-developed productive forces could potentially feed, clothe and house more and more people. Which, on its own, is correct. Galton instead believed in rational planning... not of the economy and productive forces, though, but instead in, like, state-mandated sterilization.

All this disagreement among people who are working to

ultimately preserve the same world order... It is almost as if said world order is, itself, contradictory and must be supported by a series of conflicting ideologies taking precedence in a cycle based on what sounds plausible at the time...

Galton's work was the synthesis of the previous decades, informed by everything we've discussed, and it was called... eugenics.

So, a series of thinkers took influence from Malthus's work and came to their own conclusions, but can we characterize this line of thought as specifically Malthusian? Is it reasonable to connect the concerns of overpopulation and eugenics? If only Malthus had used the word "eugenics" in his text somewhere.

I Ctrl+F'd it and it seems like he didn't. And, as all modern thinkers know, that is the only way to figure out what historical figures thought. He also didn't say the n-word so that means this ideology didn't affect black people. This is how stuff works.

Man that sucks, Malthus never said the word "eugenics" so that means he never talked about it. Guess there's no connections!

Or maybe... it means that Malthus liked eugenics before it was cool, way before Galton even came up with the word. He mused about the possibility of what we have come to call eugenics in 1798's *An Essay on the Principle of Population:*

> By an attention to breed, a certain degree of improvement, similar to that among animals, might take place among men. Whether intellect could be communicated may be a matter of doubt; but size, strength, beauty, complexion, and perhaps longevity are in a degree transmissible...

> As the human race, however, could not be improved in this way without condemning all the bad specimens to celibacy, it is not probable that an attention to breed should ever become general.

– Robert Thomas Malthus, An Essay on The Princple of Population (1798)

What's interesting here is that he asserts the ideas we would come to know as eugenics probably wouldn't become popular. He didn't think people could be convinced that sterilizing the lessers, the "bad specimens," was acceptable. This is a rare time I wish Malthus was right, but again, he was wrong.

But that's not really a program, though. Yeah, it's certainly the basic principle behind eugenics, but if only someone could connect Malthus with Galton! A smoking gun that connected this entire sequence of events and made it make perfect sense!

Alas, history rarely affords us such a simple answer. EXCEPT RIGHT HERE. RIGHT HERE IT AFFORDS US SUCH A SIMPLE ANSWER!

In his 1868 book *Hereditary Genius*, Galton cites, reworks, and "improves upon" Malthus's work, building on Malthus's concern about the procreation of what Galton called "poor quality humans," which, if the reader recalls, Malthus proposed that "by an attention to breed, a certain degree of improvement, similar to that among animals, might take place among men" as well as courting the return of the plague to cast off the "surplus population."

What Galton "fixed" was that he believed Malthus's idea of delaying marriages to later in life (at least until the population "problem" was resolved) would only be followed as a suggestion by the most sensible people, thus, only culling the "good stock" rather than the "waste people."

Because of this, Galton said this would need to be enforced on the entire population to "achieve" anything.

Galton and Malthus ultimately agreed that the biggest threat to humanity was the "lessers" reproducing without limits. Galton added that he thought the growing surplus population would continue to create misery, reinforcing "negative passions" like the "envy of the upper classes," something later repeated by anti-communists like Helmet Shoeck and Natalie Wynn.

Malthus and the ensuing so-called scientists with similar notions ultimately inspired Galton to believe that human reproduction and/or lifespan must be limited and controlled so that not only would the genetic quality of humanity increase but also that, at last, the problem of the surplus population could be solved. The inferior masses whose lives are "not worth living" would cease to be if they reproduced less or not at all, and thus, the genetically superior will flourish.

This was sold as the "solution to poverty," but it kinda sounds like... genocide, doesn't it?

The poor needed to be seen as unmotivated, inferior thinkers who could never raise themselves to the position the ruling class is in – the "slave races" even more so. How perfectly does it work out for this ideology when they pointed out that the non-working members of these groups were destitute? These folks still needed to consume food, still needed resources and housing, but the mode of appropriation was not designed to accommodate them.

If "we" could just "breed out" unmotivated, inferior thinkers, society would just be really smart people who rise to the top! We would have intellectual pursuits to occupy our energy – and would be smart enough to only breed with other smart people. All we would have to do is get the surplus population to stop banging everything that moves. We need to limit the rates of reproduction of the lesser class. Also slaves aren't human, bro, they're livestock. We can make them reproduce however the hell we want.

While Galton was part of an intellectual paradigm that can be traced back to Malthus, again, Malthus wasn't the first. Just like with overpopulation concerns, we can look to Plato for an alpha version of eugenics. Yeah, apparently, everyone with bad ideas just rips off Plato's deep cuts.

As with Malthus, Plato didn't say the word "eugenics," because Francis Galton invented that word much later. However, it's the same thing as eugenics. In *Republic*, the very same book in which Plato advocates for a ruling aristocracy of philosopher-kings called The Guardians (the most superior of all intellects making the choices for everyone else) who would, among other things, regulate reproduction because there were too many people, Plato also recommended a state-run "mating program" to strengthen... (pause for effect) The Guardians! Remember the Bang Rules? That's eugenics. I replaced the word eugenics in this image with the words "bang rules" for exactly this reason!

It's not a shock that Malthus, an influential historical guy who advocated for an aristocratic ruling class, had a lot of ideological overlap with Plato, an influential historical guy who advocated for an aristocratic ruling class.

Also, between Malthus and Plato, we're two for two on claims that there are too many people in the world and insistence that we should selectively breed humans to create "a stronger race." Throw in Epstein, and it's 3/3. Agent Smith, and it's 4/4. Man! Not a list of people I would want to be known for agreeing with.

The American Eugenics Society was founded in the early 20th century. U.S. eugenicists outwardly supported restrictions on immigration from nations with what they considered to be "inferior" stock. They argued for the sterilization of "degenerate" and "unfit" individuals – the "waste people" of "the surplus population." For reference, this could mean anything from the mentally ill, the neurodivergent, and physically disabled people to promiscuous women, homosexuals, immigrants, or racial groups.

Between 1907 to 1935, all but twelve states in the US had passed or were in the process of passing forced sterilization laws. Francis Galton died in 1911 and didn't live to see his ideas implemented on such a mass level, though he was both very unpleasant and from England, so combine those two things, and it's likely he would have thought the Yanks were doing it wrong somehow anyway.

Early in 20th century America, over 20,000 people were sterilized because of their supposed genetic inferiority, which included the poor, who, if they were genetically superior, COULD NEVER HAVE BEEN POOR, RIGHT? Meritocracy means the cream rises to the top!

Now, all this forced sterilization and genetic sorting happened in the United States *and before* the Third Reich happened in Germany. And – oh boy – the bourgeoning Nazi Party loved that shit.

Part of how it got to them was by way of a book published in 1899 called *The Aryan and His Social Role*, written by a staunch advocate of Francis Galton's by the name Georges Vacher de Lapouge. With his book, we already know the drill. He classified humanity into different, hierarchized races, spanning from the Aryan white race to the "mediocre and inert" Jewish race, distinguishing them by, among other things, skull size and shape.

These guys really all just cycled through skulls, overpopulation, and breeding humans – and they all seem to feed right back into each other, don't they? It's almost like these people don't understand that the scarcity created by exclusively designating all surplus wealth created from production to a tiny class of people who own everything is a false, imposed scarcity. Or that they don't want to understand. Hmmm!

Lapouge's "anthroposociology" helped to develop these rationalizations into a racial utopianism, a world where The Guardians, The Aristocracy, or whatever ruling class was seen as a race that was genetically predisposed to rule over everyone else. Because "anthroposociology" was regarded as a science, because all of this nonsense was regarded as a science, it could be employed as propaganda for the Nazis – one that "normal people" in Germany just had to "trust." Because we always "trust the science," don't we?

Lapouge himself had more specific aims with his work:

First, he wished to end the trade unionists, whom he and his followers considered to be "degenerate" because their intention was to improve the conditions of the working poor. Remember, according to these people, the poor were "genetically inferior waste people" and contributed to the "surplus population."

Second, he wanted to prevent social conflict by establishing a fixed, hierarchical social order with "well-bred," intelligent, elite individuals at the top and a disposable mass of people at the bottom. Except for his version of the Bang Rules became way more militarized than any before.

Vacher de Lapouge's work became one of the leading "scientific" justifications for the Third Reich[6].

Adolf Hitler detailed his belief in and enthusiasm for eugenics in 1925's *Mein Kampf* and as he took power, he implemented eugenic sterilization legislation that had been pioneered here in the United States. Legislation that I have detailed specific connection, through an institutionalized lineage of thought, to Robert Thomas Malthus, overpopulation genius.

I contend that the ultimate conclusion of overpopulation logic is some form of eugenics-justified genocide. The Nazis didn't call it the "Final Solution" for no reason; it seems like an unavoidable practice if we agree that the problems of human sustainability are anchored to there being too many people.

Whereas I oppose the idea that humanity is the problem. That's the whole point; I specifically think that humanity is a good thing.

So the US state was doing some form of eugenics-justified genocide before the Nazis were, and those guys took inspiration from it. This is ironic as would-be targets of such a program founded the US: "waste people" sent to America, Great Britain's continent-sized workhouse. But in the need to justify their ruling class, they came to act just like the ruling class they won their freedom from.

It's like there's some kind of cycle that perpetuates as long as there is an owning/ruling class that needs to explain why it owns everything and rules over everyone.

When one says "the world is overpopulated," one implies "there should be fewer people," which brings up *procedural* questions. Who do we get rid of? How do we do it? Who do we stop from having kids? Who do we deprive of food, water, resources, and healthcare to shorten lifespan? Do we just kill people? If so, who? Again, some form of Eugenics is the logical conclusion.

And it seems like the most prominent examples of that have gone pretty bad. Is there a way to go about eugenics that's... good? Wasn't Planned Parenthood founded by someone who supported eugenics? Isn't Planned Parenthood good? Don't "right-wing" people use that to try to take down Planned Parenthood?

Well, yeah. Margaret Sanger did have those views – and conservatives do use them to try to take down Planned Parenthood. Sanger advocated for eugenics in her 1920 book, *Woman and the New Race*. Hell of a title.

But here's my problem: over the years, Sanger's supporters have NOT been saying, "in a world built on economic coercion, people without the means to raise children should, as a bare minimum, have autonomy over their lives and bodies in the reproductive realm."

In fact, very few abortion advocates with large, mainstream platforms talk about economic coercion *at all* because so much of the reproductive rights sector is controlled by non-profits, which are funded by the people *doing the economic coercion*. And Sanger's supporters often try to ignore her views or even defend her as if her eugenics was somehow different or irrelevant.

So what does Margaret Sanger's good, PROGRESSIVE

eugenics want to do with "unfit" people that places it in opposition to the evil, Nazi eugenics? Rather than killing people, Sanger-style eugenicists want to stop "unfit" people from having kids, so that meant advocating a program that places reproductive authority in the hands of the bourgeois state and begs the rulers to only let the "fit" reproduce.

Merciful Marge wrote the following in her 1922 book *The Pivot of Civilization*, "Chapter 4: The Fertility of the Feeble-Minded"

> *"We do not believe that the community could or should send to the lethal chamber the defective progeny resulting from irresponsible and unintelligent breeding.*
>
> *But modern society, which has respected the personal liberty of the individual only in regard to the unrestricted and irresponsible bringing into the world of filth and poverty – an overcrowding procession of infants"*
>
> *"We prefer the policy of immediate sterilization, of making sure that parenthood is absolutely prohibited to the feeble-minded."*
>
> *- Margaret Sanger, The Pivot of Civilization (1922)*

Maggie really does an incredible job separating herself from the Nazis here; we don't KILL the "defective progeny," which in the same chapter she calls "the moron class," we spay or neuter them, like our pets!

History is full of turds, and I think that attempting to shine those turds indicates very strange priorities – often the preservation of the ruling class's power, intentional or not. Sanger is one of those turds. In a few notable ways, her actions had a positive impact, but everything that stuck eventually rejected her ideas and went a different route.

Margaret Sanger started as a socialist attempting to stop deaths from self-induced abortions by women – a noble cause. Sanger was an active member of the IWW in New York up until she fled the US in 1914 to dodge criminal charges for distributing birth control and whatever consequence one might face for calling for the assassination of John D. Rockafeller. So far, she sounds kind of badass, right?

In her 1939 autobiography, she wrote:

> *My own personal feelings drew me towards the individualist, anarchist philosophy... it seemed to me necessary to approach the ideal by way of Socialism."*
>
> -Margaret Sanger, The Autobiography of Margaret Sanger (1938)

In her 1914 exile, she went to England and fell in with the British Malthusian League founder Charles Vickery Drysdale (widely considered to be the Founding Father of Neo-Malthusianism). After a while, that Malthusian brain rot set in, and she became very concerned with overpopulation, abandoned socialism (which she was eventually going to do anyway because it was by way of "individualist, anarchist philosophy"), chaired a few International Neo-Malthusian Conferences, turned her ass around, made up with Rockafeller and began accepting funding from him to limit birthrates among working-class – er, liberate women?

Sanger published articles and papers on the subject with titles like "Preparing For The World Crisis" in which she advocated a moratorium on reproduction – no one having any kids – for five years so the population could "right itself." She also traveled around, giving speeches on things like "Overpopulation and War."

It shouldn't astound anyone that a so-called "pro-life" movement formed against a movement saying anything even remotely close to "we need to reduce the human population." This is against human life specifically, and there is no other way to put it!

The "pro-life" label makes a lot more sense knowing it is a reaction to a literal anti-life position. As with numerous other reactionary movements, this one perceived progressives doing something genuinely messed up, and due to their lack of class consciousness, just assumed that the exact opposite ideological position was the morally righteous one. That developed into the deeply reactionary battle we still see going on today over what is moral and what isn't – and what (and how much of it) the ruling class should permit.

So the logic goes, "there are too many people, and we need to get rid of a bunch of them. How do we figure out whom we get rid of?"

At the very least, eugenicists consider neurodivergence and disability to be traits of the "unfit." Racial dynamics often come into play, too, as did for Sanger. She began speaking at KKK rallies and brought Lothrop Stoddard, a white supremacist KKK member whose work was a direct influence on Nazi Germany's race law, onto her American Birth Control League's National Board.

Just for reference, the American Birth Control League became Planned Parenthood later.

There's also a ton of controversy every time anyone brings up "the Negro Project." At the time, it was claimed to be about agency and, to be frank, in an economically coercive society, having control over one's reproductive system is ultimately an issue of agency. But can we please not forget that one of the biggest goals of any of the overpopulation solutions, eugenics included, is to "reduce poverty?"

Even now, after Planned Parenthood has denounced Sanger's belief in eugenics, they simply say that Sanger "lost control of the project" and left it at that, doing everything possible to ignore the problem of a eugenicist who wanted to reduce the human population by getting rid of "the lessers," who appointed a KKK member to the board of her organization, operating specifically in black communities. If someone that outwardly pro-eugenics was doing the same thing today, I don't think people would consider it innocent.

Thing is, even if she didn't like the warmongering Nazi eugenics, she still considered the working class, with special attention for neurodivergent people, disabled people, and other specific groups "the defective progeny resulting from irresponsible and unintelligent breeding." I dare the reader to repeat that as their own opinion on Twitter. Care to imagine how that would go? Calling black people or autistic people, people with OCD or whatever "defective progeny" and calling to "breed them out?" Ooof.

Funny thing, a lot of the people who would get obsessively angry would be the same people who would defend Sanger if one started talking about the Negro Project. Honestly, the whole issue is so fucking frustrating.

Some suggest that Sanger pretended to have these views to popularize contraception, which I don't care to entertain. For one, she has been dead since 1966. I can't shoot off an email to ask her, "hey did you lie about all that shit you said and wrote to try to trick people into giving women freedom?"

Also, doesn't that sound silly? Like "she's the hero women deserve, but not the one they need right now. So, we'll hunt her, because she can take it." So Sanger the Dark Knight for reproductive rights? No. That's fucking stupid.

Is there a "good" version of eugenics in terms of practical application? Not in my view. Eugenics is ultimately about establishing who is "unfit genetic stock" and either "breeding them out" or committing outright genocide.

Poverty doesn't happen because the poor are lazy, dumb assholes who have a lot of unprotected sex because they don't know how to do anything else with their free time. It happens because there is a system set up which concentrates capital among a specific class of owners, and as it concentrates, more people have less wealth.

Some form of what people call "body autonomy" in a world that, as I said, economically coerces essentially every decision a person can make, particularly punishing single expectant parents with lower incomes, is an important cause on its own. It confuses me why people still bother defending Margaret Sanger, who, at best, complicates that fight for reproductive rights and, at worst, functions as an ideological foot in the door for eugenics.

For that matter, why indulge the Malthusian overpopulation argument at all if it does the same?

> *The old, ignorant Malthusian notions of absolute "overpopulation," or the modern lugubrious chants of birth control as the necessary solution of poverty, are thus abundantly exploded by facts. It is worth noting that this reactionary propaganda is still maintained, not only in clerical and conservative quarters but also by the would-be "progressive[s]."*
>
> *Rajani Palme Dutt, Facism and Social Revolution (1934)*

Eugenics tries to answer the question, "who should be allowed to exist?" This is a question no one would be asking if they thought that there was enough space or resources that everyone could live without impeding anyone else. This is a convenient question for the owning class because it's one that directs people away from the contradictions of class society *and towards each other*.

The proposed answer to "who should be allowed to exist" is generally not "poor people" or "various minority groups," though. The ideologies that have sprung forth both directly and indirectly from Malthusianism have been used to explain lots of unsavory views, from why caring for any poor people is actually bad to why races of people should be enslaved to why freed slaves are consigned to the lower rungs of the working class to why The Third Reich should be in power.

And that's all monstrous, horrific shit. We know that. But why can't we seem to get past it? How do overpopulation and eugenics transform themselves to continually dissuade scrutiny of the ruling class's unaccountable influence and control over every aspect of modern life? How do they get sublated (absorbed, integrated) into the more mainstream, less genocidal-sounding capitalist ideologies?

What if it's not about there being too many people but instead about there not being enough resources? And what if, instead of sterilizing them and killing them, we just make it so that the poor don't have access to those resources?

And... What if everyone believed it was vital — even unavoidable — to do this because of impending ecological collapse?

Should we... Degrow?

PART 3: DEGROWTH

The current world order of Imperial-Stage Capitalism was best described by Vladimir Lenin: the capitalist class rules through the consolidation and unification of monopoly industrial and finance capital.

Though not everyone knows it by the name "Imperial-Stage Capitalism" (some know it as "globalism"), it's fairly normal for everyday people to dislike this world order. Recent polling by Axios shows that over 40% of American people are aware of what it's like to at least some extent in existing socialist countries, like China, Cuba, Vietnam, or Nicaragua, and would prefer to live in *something like that.*

So a lot of people want an end to capitalism, but that doesn't mean everyone has a detailed understanding of what's actually wrong with capitalism. And I'm not saying that the reader doesn't, but just role-play with me here for a second. For fun, let's say we ***don't*** know the fact that capitalism's fundamental contradiction is the act of production becoming socialized among a working class while an owning class retains the product and profit.

The primary contradiction of capitalism is the socialization of production while the product and profit remained privatized, as they did in feudal times when production was not socialized. Let's say we don't know that is the thing that creates class. Let's say we also **DON'T** know that a state is an apparatus to maintain one class's rule over everyone else or that the US state is an imperial, capitalist one and therefore acts in service of the imperial capitalists.

Let's just say all we **DO** know is that we're sick of the fat cats, and we're tired of this shit. And good for us genuinely. We should all be tired of this shit.

So... we listen to a counter-culture podcast, maybe watch some leftist YouTube videos, go to a Green Day or Against Me! concert, join some message board, Tumblr or Twitter, and someone tells us to READ BOOKCHIN! So why not read some Bookchin?

> "Growth, too, was frowned upon as a serious violation of religious and social taboos. The ideal of "limit" the classical Greek belief in the "golden mean," never entirely lost its impact on the precapitalist world. Indeed, from tribal times well into historical times, virtue was defined as a strong commitment by the individual to the community's welfare and prestige was earned by disposing of wealth in the form of gifts, not by accumulating it."

- Murray Bookchin, Remaking Society (1990)

Ohhhhhh! We've lost the morals that fortuitously put restraints on growth? No wonder things have gotten so bad, we're just greedy, gluttonous, bad people. We really should get back to the classical beliefs of ancient Greece! What were we saying earlier?

"It's not a shock that Malthus, an influential historical guy who advocated for an aristocratic ruling class, had a lot of ideological overlap with Plato, an influential historical guy who advocated for an aristocratic ruling class. Also, between Malthus and Plato, we're two for two on claims there are too many people in the world and insistence that we should selectively breed humans to create 'a stronger race.'"

Ah, right. Forgot about that. Uh... Maybe Bookchin means a different Ancient Greece? Okay, maybe not. Who cares? What does he think about CAPITALISM!?

> *"Capitalism can no more be 'persuaded' to limit growth than a human being can be 'persuaded' to stop breathing."*
>
> *- Murray Bookchin, Remaking Society (1990)*

Yes! The problem is for sure growth! Capitalism's main problem is infinite growth!

> *"Questions of growth, profit, the future of the planet, [are] no longer single issues or class issues but human and ecological issues."*
>
> *- Murray Bookchin, Remaking Society (1990)*

Finally, we're getting somewhere! The problem with capitalism isn't a **fundamental contradiction of socialized production and privatized appropriation of product and profit, thusly creating both a ruling class who unaccountably makes the choices that steer society and a structural flaw that creates a cycle of exploitation and crisis** – like I just said like a second ago when I was talking about how things actually are.

It's nothing like that! It's just that capitalism simply "demands" infinite growth and the planet has ecological limits! Maybe there just needs to be less consumption overall, right?

> *The thing about growth is that it sounds so good. It's a powerful metaphor that's rooted deeply in our understanding of natural processes: children grow, crops grow ... and so too the economy should grow. But this framing plays on a false analogy. The natural process of growth is always finite[...]*
>
> *The capitalist economy looks nothing like this. Under capital's growth imperative, there is no horizon – no future point at which economists and politicians say we will have enough money or enough stuff. There is no end; the unquestioned assumption is that growth*

> *can and should carry on forever, for its own sake. It is astonishing, when you think about it, that the dominant belief in economics holds that no matter how rich a country has become, their GDP should keep rising, year after year, with no identifiable end point.*
>
> *- Jason Hickel, Less is More: How Degrowth Will Save the World (2020)*

The last Bookchin quote leads into this passage from a 2020 book called Less is More: How Degrowth Will Save The World. The author, Jason Hickel, who, to avoid confusion, is NOT Jackson Hinkle, printed the Bookchin quote about persuading capitalism to justify THIS argument. In fact, it's literally directly before it in the book.

Bookchin, an anarchist, viewed the earth's problems as "general interest" rather than the continued struggle between the classes of the rulers and the ruled. For an anarchist that supposedly views hierarchy as a corrupting force, this is very trusting that the ruling class will do the right thing for everyone else! Bookchin said a lot of shit like this and that's likely why Degrowth Influencers use his work.

Jason Hickel, who is himself a Degrowth Influencer, is one of many who represent their arguments as "very obviously about the economy," to the point where they even frame themselves as anti-capitalist, all while being funded and/or promoted by capitalists. Don't dare assert Hickel or whoever is pushing anything Malthusian, either, or they and/or their legion of extremely online fans will demand one reads *Less is More*, because if one thinks it's Malthusian, they're obviously misunderstanding it or misrepresenting it.

I'll talk more about Degrowth Influencers (and the capitalists that love them) in a moment. But first, more on *Less is More*. Advocates of degrowth would fervently deny any association with Malthus or overpopulation and as such, Robert

Thomas Malthus isn't mentioned even a single time in the main text (so don't even try to Ctrl+F it!).

There is a mention – *one* mention – of Malthus in the citations list. There's a book cited called *Limits: Why Malthus was Wrong and Why Environmentalists Should Care*.

Which sounds great to someone who understands Malthusianism, because Malthus IS wrong! I mean, there's only one reason why one would say "Malthus is wrong," and it's because of the main thing people know about Malthus, right? Otherwise, we wouldn't even be talking about Malthus, right? So the author Giorgos Kallis must be arguing against overpopulation, right? RIGHT!?

The way Kallis considers Malthus wrong, however, is that he believes Malthus was (wait for it) pro-growth. And not just pro-growth but pro-capitalism. The guy who said that there's going to be so many people there's a disaster because the population is growing out of control due to industrialization, and who continually advocated for the subordination of the capitalists to the aristocracy, was a pro-growth, pro-capital guy, I guess.

This is actually a fairly similar problem that founder of eugenics Francis Galton had with scientific racist and advocate of free market capitalism, Herbert Spencer – as well as, ultimately, Malthus himself as well.

Again, Spencer thought it was fine that any disaster might happen because white people would be able to take advantage of it, and that's why the free market is good. And while Galton didn't dislike Malthus, he did dislike Malthus's pessimism, perceiving Malthus to resign humanity to its fate, despite his valiant criticism.

But where Galton believed that policy could be implemented from a place of authority (and was proven right, on that single point, by the Nazis, who killed a bunch of people from a place of authority), Kallis argues for environmentalists to push the idea of humanity's "voluntary" limiting of itself, noting "our world is limited because our wants are unlimited."

Here, we see instances of the two ways capitalism preserves itself: in Galton's take on Malthus, brute force; in Kallis's, ideology is used to coerce compliance.

As we consider either, we must remember Karl Marx calling Malthus's work simply "a lampoon meant to defend the feudal aristocracy;" I see it as doubtful that Marx saw Malthus as anything above cynical, calling him a "professional sycophant" among other colorful insults.

We can see the coercive element in what Kallis neglects to mention. Strangely absent from his "Malthus was wrong" argument is Malthus's main conceit of overpopulation, the one for which he is primarily known, the common thread that ran through race science and eugenics. In fact, in promoting the idea that humanity should "voluntarily limit" itself, Kallis employs a rhetoric of recreational sex, using something that sounds completely awesome to casually impart to the reader that humans should reproduce less, thus reducing the population.

Not a good book to cite if the goal is trying to tell people the book isn't about overpopulation or lowering the population. Especially considering the point of Kallis's book is to inculcate in the reader a "voluntary" model of self-limitation ideology.

But the honest thing to do is to give *Less is More* a chance, right?

> *To imagine that we can continue expanding the global economy indefinitely is to disavow the most obvious truths about our planet's ecological limits. This realization first struck home in 1972, when a team of scientists at MIT published a groundbreaking report titled Limits to Growth. The report outlined findings from the team's cutting-edge work using a powerful computer model called World 3, which was designed to analyze complex ecological, social and economic data from 1900 to 1970, and to predict what would happen*

to our world in twelve different scenarios by the end of the twenty-first century.

The results were striking. [...] 'The most probable result,' they wrote, somewhat ominously, 'will be a rather sudden and uncontrollable decline in both population and industrial capacity.'

It touched a nerve. Limits to Growth exploded onto the scene and became one of the best-selling environmental titles in history, tapping into the countercultural ethos that prevailed in the wake of the youth rebellions of 1968.

But then the backlash came – and it came with overwhelming force. The report was denounced in the pages of the Economist, Foreign Affairs, Forbes and the New York Times, and big-name economists came out railing against it. They said that the model was too simplistic. [That] it didn't account for the seemingly limitless innovation of which capitalism is capable.

- Jason Hickel, Less is More: How Degrowth Will Save the World (2020)

Hickel neglects to mention... well, a lot here. Firstly, the backlash was more about the Malthusian overpopulation arguments that permeate *The Limits to Growth* than the methodology. Hickel does the same thing as Kiallis here and ignores the most obvious problem and addresses a completely different argument than the one that matters.

Second, *The Limits To Growth* wasn't just the result of some cool dudes having a good time playing with computers at MIT. An organization called The Club of Rome commissioned the report[7]. The Club of Rome was a think tank founded by Aurelio Peccei, an Italian business executive active in promoting international trade, with the backing of Giovanni Agnelli, the Chairman of Fiat, commander of literally 4% of Italy's GDP at the time, and a David Rockefeller appointee to the International Advisory Committee of Chase Manhattan Bank.

Also a co-founder of The Club of Rome: David Rockefeller. Basically, The Club was one hell of an advocate for international finance capital that was willing to use population rhetoric to create concern that would back its policy ideas.

Reminds me of a recent Club. A Good Club.

The fundamental point of the organization was to persuade governments in imperial nations that a new regulatory body should be formed internationally by finance capitalists because they knew the world's business better than anyone else. Clearly, it was beyond the bodies of states' capabilities.

Limits was ultimately going to be propaganda (or, as Peccei put it, "tools of communication and conviction") in service of that agenda. Don't take my word for it, though, Peccei specifically said so in his 1969 book *The Chasm Ahead:*

> *Nowadays all peoples are awed and fascinated by the new technologies they do not understand, far less dominate. In my opinion therefore, they are prepared for quite a number of years, and on condition, to recognise a new world moderator or even a new authority, set up by those who master the esoteric technologies... even if it is a far away, [a] supernational, non-personalised and vicarious authority.*
>
> – Aurelio Peccei, The Chasm Ahead (1969)

Second, even the Club of Rome – still primarily under the control of Aurelio Peccei – disavowed the study shortly after it was published, likely because Peccei's goal was not to advocate against industry but for the creation of new regulatory bodies directly representing the will of international finance capital.

In my perception, Malthusianism was likely convenient to Peccei's motives, and his goal was not specifically spreading Neo-Malthusianism. But he died in 1984. Alexander King, a scientist and another co-founder of the Club, outlived him. Spreading Neo-Malthusianism *clearly* was his goal as he took over as President of the Club of Rome.

> *The common enemy of humanity is Man.*
>
> *In searching for a common enemy against whom we can unite, we came up with the idea that pollution, the threat of global warming, water shortages, famine, and the like, would fit the bill. In their totality and their interactions, these phenomena do constitute a common threat which must be confronted by everyone together. But in designating these dangers as the enemy, we fall into the trap [of] namely mistaking symptoms for causes. All these dangers are caused by human intervention in natural processes, and it is only through changed attitudes and behavior that they can be overcome. The real enemy then is humanity itself.*
>
> - Alexander King, The First Global Revolution (1991)

So, though Club of Rome's Neo-Malthusianism is presently its primary purpose, at the beginning, it was seemingly a means to popularize and justify Aurelio Peccei's agenda. And among the world's elite, *The Limits to Growth* was *very* popular (and still is). But everyone else? Not so much.

> *The best-known doomsday forecast in the last few decades was The Limits to Growth [which] has been so thoroughly and universally criticized as neither valid nor scientific that it is not worthwhile to devote time or space to refuting its every detail. Even more damning, just four years after publication it was disavowed by its sponsors, the Club of Rome. The Club said that the conclusions of that first report are not correct and that*

> *they purposely misled the public in order to "awaken" public concern.*
>
> *- Julian Simon, "Why Are Forecasts So Often Wrong?" from "The Ultimate Resource 2," (1981, updated 1996)*

Third, The Limits to Growth should sound familiar. I mentioned it in part one along with the author of the quote I just read, Julian Simon. Simon spent several decades debunking the revived neo-Malthusian movement that exploded in popularity in The Limits to Growth's blast radius[8].

Again, this is a person who I heavily disagree with, economically speaking; a pro-market liberal of the Chicago tradition (meaning hard neoliberal) who served as an environmental economics professor at the University of Illinois.

According to Simon, he started out a "card-carrying anti-growth, anti-population zealot," but after he actually did an analysis of the numbers, "the data did not support that original belief." He criticized Malthusians' "lack of historical perspective" and tendency to think of resources as "autonomous of human productive and creative forces as if they were independent of human action, and impervious to transformation through technology, choice, and inventiveness." He called this perspective that of a "closed system," one dominated by the concept of fixedness and finiteness, which "gives the illusion of easy, calculable, and uncontroversial 'scientific' results"[9].

On this subject alone, Simon ends up sounding more like a Marxist than any other Chicago School of Economics proponent *I've* seen. Unfortunately, this is limited (no pun intended) to Simon's criticism of Malthusianism and the Limits movement.

Still, the limited perspective of a "closed system" calls to mind the previously mentioned point Caleb Maupin made about how "the way human beings interact with resources has constantly been in a state of change." Maupin is, himself, directly calling on the dialectical materialism put forward by Karl Marx and Fredrich Engels – a method that acknowledges the world we live in as deeply connected and constantly in a state of flux:

> *In the contemplation of individual things, [a metaphysical analysis] forgets the connection between them; in the contemplation of their existence, it forgets the beginning and end of that existence; of their repose, it forgets their motion.*
>
> -Friedrich Engels, Socialism: Utopian and Scientific (1880)

One of the most important insights Simon raises in debunking *The Limits To Growth* (and other Neo-Malthusian works one might hear Degrowth Influencers quote, despite being "totally not Malthusians") is that technical knowledge does not occur spontaneously.

We've all heard stories about some invention randomly exploding into some guy's head while he's eating an apple or taking a shit or something, but that isn't *really* how it happens. Simon, just as Marx and Engels likely would have, asserted that the link between needs, social conditions, and knowledge was either misunderstood or totally neglected in these kinds of abstract, predictive models.

This is, essentially, the same argument Engels posits against restricting one's investigation to metaphysics:

> *This method of work has also left us as legacy the habit of observing natural objects and processes in isolation, apart from their connection with the vast whole; of observing them in repose, not in motion; as constraints, not as essentially variables; in their death, not in their life.*
>
> - Fredrich Engels, Socialism: Utopian and Scientific

And, just like with the Malthus curve, we can also easily

debunk *Limits* using numbers.

With the data from *Limits*, and assuming consumption stayed at the same level (neither grew nor degrew) one could predict that the world would run out of gold by 1981, mercury by 1985, tin by 1987, zinc by 1990, petroleum by 1992, and copper, lead and natural gas by 1993. Essentially, all of those reserves would have been entirely depleted before the year 2000.

Gold is a vital component in printed circuit boards. So computers, electronics, etc. They all have gold in them. If *Limits* had been correct, we would have had to stop making printed circuit boards in 1981. So the NES, the SNES, the Sega Genesis, the Sega Saturn, The Sony Playstation, the Nintendo 64, and, ultimately, the Sega Dreamcast, the zenith, the absolute peak of video gaming, and then every inferior gaming console that come after it would have never happened and Gamers would have never become the Most Oppressed Group of People.

But *Limits* was not correct, which is just too bad for Gamers!

The USGS reports that 50 years later, the rate of consumption has vastly increased. And rather than see depletion in these reserves of so-called nonrenewable resources, they have grown significantly.

Now, the numbers aren't what matters to me so much as the flawed mechanics. Numbers like this only show us as close to an objective measurement of the *result* as we can get. However, showing that the asserted mechanics predict the exact opposite of what actually happened does really help make the point.

So, with all this in mind, it is interesting to find that Jason Hickel finds *The Limits to Growth* so "groundbreaking" and that the legions of degrowth fans who demand we read *Less is More* to prove that "degrowth is not Malthusian" seem entirely unaware of – and unable to learn about – any of this context surrounding the report itself.

The primary arguments of *Less is More*, coincidentally, echo specific components of Simon's critique of *Limits*. In *Less is More*, Hickel rejects technological solutions to resource usage and carbon emissions; from damning nuclear with faint praise to outright rejecting carbon capture. Instead, he puts forward that we must minimize consumption so we can switch to 100% intermittent renewable energy – which couldn't consistently keep up with peak demand hours.

The problem is that the only way to have constant energy with intermittent renewables is to have a baseload from another source of energy. Nuclear would be a great one, but these people at times totally ignore nuclear while at other times arguing against it in one way or another. Hydroelectric is pretty good where it can be done, but not everyone lives near the Hoover Dam, either.

This leaves us with only one other truly scalable option: natural gas. Which these people also argue against. Natural gas is neither "renewable" or "clean," which defeats their stated purpose.

So what about batteries, which would supposedly mitigate or end the issues with intermittents? Well… batteries are not actually an option. An average household of 2.5 people consumes 10,715 kWh per year[10]. Residential energy consumption is about 21% of total power consumption[11], so for a city of 1 million people, that would average out to 395,257 houses. So… 395,257 * 10.715 kWh = 4,235,178 (four million two hundred thirty-five thousand, one hundred seventy-eight) kWh total consumption over a year, for a city of 1 million.

At the current cost of $345 per kWh, according to the National Renewable Energy Laboratory's yearly report, it would cost $1,461,136,670 (one billion, four hundred sixty one million, one hundred thirty-six thousand, six hundred seventy) to store enough energy for 1 million people doing absolutely nothing business related, hanging out at home. There's 8 billion people on earth, and a thousand million in every billion. So, scale up that cost and we have $11,689,093,360,000 (eleven trillion, six hundred eighty nine billion, ninety three million, three hundred sixty thousand) for people to Netflix and chill, or whatever. Residential power use being about 1/5 total power use, let's toss the other 4/5ths in and... wow it would cost something like $58,445,466,800,000 (fifty eight trillion, four hundred forty five billion, four hundred sixty six million, eight hundred thousand).

Humanity only has... $40 trillion total currently. Ooof. There's not even enough total capital in the world to do that. Never mind how bad mining that much lithium would be for the planet and indigenous people, whom all these people claim they care about so dearly.

So... are we doing 100% renewables? Are we cutting off power intermittently? Because that absolutely will reduce the population.

Still, is there any way to really pin down what the motive actually is? Maybe we need to think about the mode to determine the motive. Maybe that is really how we really get to the bottom of Degrowth.

Less is More, like *Limits* and Malthus himself (with his crude, imaginary model depicted in the Malthus Curve) insist that humanity's relationship to the earth's resources is fixed and finite – not something that continually changes, though Hickel sometimes says exactly the opposite. To claim that everything is connected and that humans should co-exist with nature while also claiming that we should only develop intermittent energy solutions that ultimately reduce our capacity to sustain human life should produce cognitive dissonance with any critical examination. It's a transition from a pro-humanity view of human existence to one of the supremacy of nature.

That last one is key, too: with an overpopulation or degrowth paradigm, we're no longer seen as a species existing and evolving as part of Earth's ecosystem. We are somehow uniquely above animals while also uniquely below them. We have The Hunger! Greed and gluttony are part of human nature and we must control it! Humanity is a cancer; a virus. Basically, the same viewpoint the bad guy in *The Matrix* had.

Simon critically noted these issues in *Limits* and the ensuing movement around it decades ago.

Hickel tiptoes around this by seemingly embracing a Bernie Sanders approach to policy; claiming we must redistribute to create "radical abundance." However, this is where we find the "Limits to Social Democracy." The Nordic Model depends both directly and indirectly on the rampant exploitation of imperialized nations to create an image of abundance in its core states – an ultimately temporary one, at that.

This requires rapid, often unregulated industrialization in the so-called third world under the direction of the imperial capitalist class, doing the opposite of what Hickel claims. Not that industrialization is itself bad – in fact, for genuine radical abundance, humanity must do so. Hickel even concedes this in his paper *Degrowth: a Theory of Radical Abundance*, noting the so-called "third world" must be allowed to grow. Curiously, though, he still imposes percentages of these countries' GDP that must be reduced.

So how do these nations grow while at the same time degrowing all at the behest of imperial nations which extract their value in order to do their so-called redistribution that somehow makes degrowth "not austerity." It's all contradictory and wouldn't work.

Growth is the motive Degrowth advocates assert capitalism is driven by – and this is a key error. The assertion here is that motive dictates mode, that because "humans are greedy," we engage in the kind of production our species engages in. However, the exact opposite is correct; the ruling motive is entirely downstream of the ruling interest. The current, capitalist mode of production contains a fundamental flaw that creates classes of haves and have-nots, rulers and ruled. The motive of the ruling class, the one that makes the decisions about what happens in society, is to preserve the concentration of capital through profit and/or ownership.

Any change in motive Degrowth advocates propose without a change in interest, this would merely be symbolic. But Hickel doesn't advocate for any change in ownership, so there's no reason to believe Degrowth would be done in a way that is anything but exploitative, because if ownership isn't addressed, exploitation isn't addressed.

Even if his so-called "Theory of Radical Abundance" agenda was implemented, it wouldn't work. It would *necessarily* create more inequality between the more and less developed countries, more inequality within the imperial core countries, and cause more human suffering everywhere.

To Hickel, Humanity's supposed relationship to the earth is incoherent; we're above other species but uniquely destructive – we're incredibly smart and adaptable but also can't stop expanding and consuming. As with Bookchin, this lends to thinking of capitalism as an environmental crisis we must convince the capitalists to reorganize and impose controls so at least a portion of humanity can continue, when in reality it is, like all previous societal struggles, one of class.

But what does this mean in terms of population?

Again, most degrowth people vehemently deny they are talking about the population but rather the population's consumption, but the trick is that human consumption and human existence are intrinsically linked.

Jason Hickel
@jasonhickel

It's not overpopulation that causes climate change, it's overconsumption | Fred Pearce
gu.com/p/4xyjn/stw
9:07 AM · 07 Dec 15 · Twitter Web Client

"Overconsumption" is when the usage of resources exceeds the sustainable pace and capacity of the ecosystem. "Overpopulation" is when the usage of resources exceeds the sustainable pace and capacity of the ecosystem... because there are too many people!

Obviously very different! But *who* is being asked to consume less? Is it lions? I mean they do eat a lot but no. That's not who's being asked to consume less.

It's humans, because we are uniquely able to have morals but also have a greedy nature we must resist to Make The Sacrifice™ and Return To Monke. It's not austerity and it's not population control either! It's just reducing energy consumption! After all, energy production is evil, right? Something that greedy capitalists do to... make smoke. And money, too. Proft, ew, yuck! So let's maybe think about how we, personally, can consume less:

The concept of the "carbon footprint" was created by marketers at BP to shift responsibility for pollution away from them and onto the consumer. The way they manage this is by saying they provide a product and demand is why. "You demand it, we supply it." And that's pretty much how we're supposed to believe everything in a free market economy works.

While at some point it might have, it definitely hasn't since I've been alive. I mean, if production was dictated entirely by what was demanded, why would advertising with the purpose of creating demand for an already-existing product (all contemporary advertising) even exist?

Further, a number of studies from the likes of Harvard and Yale in recent years have indicated pretty much all of the messaging from all of the oil companies is this same demand-focused messaging and any problem that we *would* have with them, they want us to have with our neighbor that doesn't recycle.

> **CONSUMERISM LABEL URGED**
>
> CHEYENNE, Wyo. (AP) — John S. Bugas, a vice president of Ford Motor Co., Saturday night proposed "consumerism" be substituted for "capitalism" as a description of the American economy.
>
> "The term 'consumerism'," he said, "would pin the tag where it actually belongs — on Mr. Consumer, the real boss and beneficiary of the American system.
>
> "It would pull the rug right out from under our unfriendly critics who have blasted away so long and loud at capitalism. Somehow, I just can't picture them shouting: 'Down with the consumers!'"

"It's your fault for needing so much of our horrible, planet-killing energy! Oh, we do so regret having to *supply* this energy you *demand* of us with such fervor!"

In 1955, John S. Bugas, then-VP of Ford Motor Company, proposed the term "consumerism" as a substitute for "capitalism" to "better describe the American economy" and illustrate what he called "consumer sovereignty" in a defense of the capitalist market economy.

According to Bugas, "the term consumerism would pin the tag where it actually belongs – on Mr. Consumer, the real boss and beneficiary of the American system. It would pull the rug right out from under our unfriendly critics who have blasted away so long and so loud at capitalism – somehow I just can't picture them shouting down with the consumers!"

He was wrong about that last part. Look at critics of

capitalism in the neoliberal era... Look at Breadtube. Look at NPOs and NGOs. All of their rhetoric is pinned on Mr. Consumer and his morals. The way Bugas was wrong was actually quite beneficial for capital, though; people who noticed environmental problems but don't incorporate material analysis, but rather just want to moralize a better future into existence – putting a demand out and waiting for it to be supplied.

As "2015 Jason Hickel" said, it's not overpopulation that causes climate change, it's over overconsumption. But what if someone counters by articulating an argument about consumerism not actually driving capitalism?

That makes it sound like overconsumption is merely a symptom of a growth-addicted system. Capital is required to find ways to get people to consume more and more. It is a structural imperative. We have to target the real causes here. It's like "2018 Jason Hickel" says:

Jason Hickel ✓
@jasonhickel

People. Overconsumption is not the problem. It is merely a symptom of a growth-addicted system. Capital is required to find ways to get people to consume more and more. It is a structural imperative. We have to target the real causes here.

10:04 AM · 21 Oct 18 · Twitter Web App

75 Retweets   **7** Quote Tweets   **262** Likes

It sounds like we just need a new set of rules. I wonder if Jason Hickel might have some rules we can follow. HMMMMM.

The Rules was a "global network of activists" and/or a marketing firm. I'm not really 100% sure what the difference is these days, but it's especially hard to nail down exactly what The Rules was because it never really put forward a firm description of itself.

Formally, "The Rules" no longer exists. Still, its work continues as a "co-op consultancy focused on data-driven research and strategy services for systems change" called Culture Hack Labs, whose mission statement is to "expose, disrupt and shift cultural assumptions to create new narrative spaces for possibility, hope and justice."

Hickel was an influencer on hand for the organization, co-founded by Martin Kirk (another Degrowth Influencer, although one who seems content to be in the background a little bit more) and others. Hickel apparently helped sculpt the foundational content of the org and continued to be a public representative for them throughout its existence, writing articles with Martin Kirk – like one from 2017 with *Fast Company* entitled "Don't Be Scared About The End Of Capitalism — Be Excited To Build What Comes Next"[12].

In this article, Hickel and Kirk call existing socialist states "an unmitigated disaster" while praising Scandinavian states – ironic because imperialism's fast-burning through of resources funds those states. They claim their degrowth ideas of "reducing GDP" are "informed by up-to-date science" and that they "focus on regenerating rather than simply using up our planet's resources." They talk about "how to live well within ecological boundaries" and that we should "draw on indigenous knowledge and lore about how to stay in balance with nature," an incredibly reductive, fetishist statement that leans on the stereotype of the noble savage.

Most importantly, Hickel and Kirk, representing The Rules at the time, tell us about how "we are overshooting Earth's carrying capacity by a crushing 64% each year, in terms of our resource use and greenhouse gas emissions."

Now, if one were to ask them if they were talking about overpopulation, they would say "no." But consider that the reason the population is 8 billion, rather than a fraction of that, is industrialization – the mechanizing of resource use. Population growth did not pick up speed until the industrial revolution when methods of energy output that could sustain larger populations were invented.

Making drastic reductions in energy output will kill people. Temporary, incidental power outages in Texas killed hundreds in the winter of 2021. This was due to problems, not due to switching to a source of energy output that generates much less electricity.

So... where did the money The Rules used to promote the degrowth agenda come from? Well, Hickel and Kirk wowed investors like Ian McClelland of "The HIF," which is the "innovation funding" arm of Elrha, a weirdly named "global charity that finds solutions to complex humanitarian problems through research and innovation."

Aside from whatever they bilked out of She-Ra or whatever, The Rules also took a lot of money from NoVo (via a pass-through organization called Tides Foundation) over the years.

NoVo money comes from Peter Buffett, son of Warren Buffett. Remember The Club? Not the Rome one, the billionaire capitalists that got people all riled about overpopulation several decades ago – the Good one. The billionaire capitalists that got people all riled about overpopulation only one decade ago. Peter Buffett is the head honcho of NoVo, which is funded by his dad, Good Club member Warren Buffett.

In September of 2021, Peter posted an article to his Facebook feed written by Robert Jensen entitled "Restless and Relentless Minds: Thinking as a Species out of Context"[13].

I'll quote something of interest from this article:

> Is the current 8 billion people at the current level of aggregate consumption sustainable? Is an even larger population, continuing to consume at current levels, sustainable?

[...]

I do not believe either of those possibilities is plausible. That means that we need to act today to make possible a future with fewer and less. Fewer people consuming less.

How many fewer? How much less? Again, no one knows or can know for certain. But as we make choices today, we have to make our best guess. For purposes of starting honest conversations about public policy, I assume the answer is no more than half the current population consuming no more than half the resources used today. That likely won't be enough, but it's a start.

-Robert Jensen, Restless and Relentless Minds: Thinking as a Species out of Context" (Resilience/Post Carbon Institute)

Just to clarify, this is Peter Buffet-described "great mind" Robert Jensen calling for reducing the world population by half *as a start.* This is the Good Club member's son that funded The Rules as they told us first that overconsumption is the problem and then that overconsumption was a symptom of capitalism, which the real problem with is growth. NOT POPULATION! IT'S NOT POPULATION, GUYS!

So the money says one thing (in pretty plain terms) and the guy saying stuff on behalf of the money is claiming that's not really the case. But it's always been the case, even going back to Marx debunking Malthus:

> *Malthus also wishes to see the freest possible development of capitalist production, however only insofar as the condition of this development is the poverty of [...] the working class, [which he wants] to adapt itself to the consumption needs of the*

> *aristocracy and its branches in State and Church, to serve as the material basis for the antiquated claims of the representatives of interests inherited from feudalism and the absolute monarchy."*
>
> - Karl Marx, Theories of Surplus Value (abridged quotation, 1863)

Marx says here that Malthus supports capitalism as long as the wealth extracted from the working class's labor is used to preserve the position and comfort of the aristocracy. This is part of Marx's overall argument that Malthus was essentially inventing ways to justify the oppression of poor people (and to indirectly rebuke the French Revolution with pseudo-science). The aristocracy, whom Malthus represented, saw itself as inherently above commoners and that they deserved to be upheld by their lessers. Malthus saw the revolution in France as the product of these lessers not knowing their place.

The world is *"just"* when the lessers stay in their tiny houses and consume less. They must expect less so the capitalist ruling class can continue to have more. They're the smarter, better people because God chose them! Accept less, lessers. *Less is More...* For thee! Not for me! These ideologies continue to crop up from time to time – although we've replaced God with an Invisible Hand.

In 1973, the year after *Limits* was published, an economist named E. F. Schumacher published a collection of essays entitled *Small is Beautiful*. The point, more or less, was to tell people that bigger *isn't* actually better. Smaller is better! Less is more!

Tiny houses, less consumption, less energy demand, less industrialization, less industry, less civilization. We should be centering indigenous voices for their inherent magical knowledge – their innate spirituality – they're so much more in tune with nature. So much nobler! So pure!

We should be living like that. We should be living on $2.50

a day! We shouldn't be trying to simply solve these problems – we clearly can't in capitalism! We should just ignore them and the capitalist ruling class will decide what to do. We should just retreat into a small cabin in the woods and write about blowing society up... Yes... writing...

"Small is Beautiful" ideology is credited to Schumacher, but as with much of this, isn't new. Again, I'll simply hand it over to our man Rajani Palme Dutt, who dismantled it in 1934.

> *An example of the popularisation by finance-capital of this reactionary propaganda in its most fantastic form may be noted [in] an article prominently published in the millionaire-owned Sunday Express under the title, "Make Way for the Small Man," denouncing the illusion of "Progress" and the failure of "mass production," and calling for the return to "the small owner" as the ideal:*
>
> **"The unit of the [self-supporting] State is the self-supporting farm with first thoughts for subsistence and only second thoughts for the market-which might be mainly next door and consist of craftsmen supplying the needs of neighboring farms.**
>
> **This simple farm-and-craft relationship is essential to the health and wealth of any civilization... We should try to recover it."**
>
> *- Sunday Express, January 15, 1933*
>
> *The finance-capitalists would be highly indignant if this infantile propaganda, which they broadcast by the most highly developed "mass-production" for the befogging of their readers, were suggested to be seriously applied to their [own] mammoth undertakings, including their mammoth Press. The*

> *preaching of monopoly-capital against monopoly is an old story.*
>
> *Return to handwork! Return to the Stone Age! Such is the final logical working out of the most advanced capitalism and Fascism."*
>
> *- Rajani Palme Dutt, Fascism and Social Revolution (1934)*

So whether it's Malthus pushing overpopulation concerns to support the aristocracy, Club of Rome pushing *The Limits to Growth* in support of international finance capital, The Schumacher Institute pushing Small is Beautiful, The Good Club pushing its agenda through philanthropy, or various Degrowth Influencers funded by the capitalist class (either directly or through NGOs/NPOs and passthrough organizations), it seems obvious who benefits. It's the ruling class.

Whether anyone who pushes overpopulation rhetoric genuinely believes or cynically employs it, I don't know. I can't say what is going on in other people's minds. But guess what? I don't actually care at this point. Every one of these people could be dead serious about believing in overpopulation being our most pressing concern or straight-up lying. This doesn't change the outcome.

So long as the capitalist mode of production exists, artificially restricting the productive forces will always eventually be necessary due to the system's inherent contradictions. A population WILL be reduced, maybe because of some insane Naziesque eugenics solution or maybe in a World War. Maybe the former causes the latter. Maybe just austerity and coercive ideology. Regardless of how it's utterly necessary.

It's not because the population is consuming too much; it's because the mode of production doesn't work. The capitalist ruling class has power because, as I said earlier, power comes from production. If one owns *production,* one owns the prospect of "if people eat," among many other things. In the imperial mode, one which grows markets until they're no longer sustainable and then degrows them, redrawing territorial boundaries, there's an inherent flaw that causes cyclical crises.

Let's talk about industrial society... and its past.

According to Marx's critique of capitalism, the things we purchase and consume are only as valuable as the amount of labor that's necessary to produce them.

Technological innovation reduces the amount of labor necessary to produce something, so according to Marx's paradigm, that thing becomes less valuable, at least in terms of what someone will pay for it at the market.

Every single time an old person says, "they don't make them like they used to." They're completely right.

Technology and technique develop; what was once handmade becomes machine-made. In this development alone, things cost less to make. When a material that takes labor to be mined and refined can be replaced with plastic, we find the further reduction of labor continues to reduce the cost.

Capitalists who get to technological innovations first are the ones who benefit most from this lowered production cost, as the new technology takes time to become adopted as the production standard. Once it does, all the would-be competing products cost the same (or nearly the same) to make. And so it costs less to buy, because not only does this process cause inflation, but people aren't going to pay a premium for something that's "not made how it used to be."

Despite this, demand for a product people *need* will likely stay the same. Even if the demand for something stays the same while less labor is required to produce it, the value of that thing goes down. Therefore the profit margin goes down, as capital must produce the same numbers of this now-cheaper thing to fulfill demand. This is called the falling rate of profit. It is baked into our economic system. It is totally unavoidable.

Because of this, it isn't actually in the interest of capitalists to fulfill demand in the long term. In fact, if capitalists continually focused on supplying to fulfill demand, they'd eventually go out of business.

Hence, degrowth.

> *"The modern development of technique and productive powers has reached a point at which the existing capitalist forms are more and more incompatible with the further development of production and utilization of technique."*
>
> *"To the modern bourgeois mind and outlook, this process of... restricting of production, in the midst of poverty, appears as a natural and self-evident necessity. Without a sense of contradiction, they proclaim... the policy of restriction of production with the same sense of obvious correctness and common sense with which they preached after the war, the policy of "increased production" as the path to prosperity.*
>
> *"Correctly they feel no contradiction since both are indispensable to the maintenance of capitalism at the present stage."*
>
> *- Rajani Palme Dutt, Fascism and Social Revolution (1934)*

Degrowth is an ideology that naturalizes and stabilizes the falling rate of profit under capitalism. It shifts the responsibility of capitalism's inability to provide for those outside the ruling class to controlling consumption habits and forcing austerity measures on the masses.

The degrowth agenda is generally built around reducing demand for energy (and therefore everything) to the point where

"100% intermittents" is a viable situation that could actually provide enough power for the total societal usage. This is, I must stress, a massive reduction in energy production and will kill people if actually implemented.

Degrowth advocates, funded and/or promoted by capital, are forcing "discourse" about a hypothetical artificial limitation of production and technological progress, and that is where the conversation stays. In the people, in congress, in the Oval Office. It's marketing, and they don't need everyone to sign on. In fact, if everyone splits into factions and argues about which ideals are better rather than what the system does and who benefits, it's easier for them to avoid any kind of accountability at all, as people who could devise a way to organize labor to subvert this are arguing about which brands buy more carbon offsets or if we should all become trad, Anarcho-Primitivist, or localist. Should we start a community garden, ban cars tomorrow, and stop having kids for five years? Are people the real virus? Should we just move to the country and eat a lot of peaches?

What is right for me, individually, as a person to do? What are the right ideals? What are the right morals? What is good and what is bad?

These are the questions this flawed system needs us to be asking when we notice there are problems rather than questions about the relationships and conditions that create this situation.

Energy companies are ultimately finance capital. According to Lenin, finance capital is monopoly industrial capital merged with bank capital (which actually doesn't require a single Jew to be involved, as some idiot is probably going to assert in either the comments or a response video). Investor-owned utilities generate roughly 80% of US energy.

With some scrutiny, we can see why energy companies are pushing windmills and solar panels. We can also see why they fund non-profits that push these types of energy paradigms and why they're happy to have degrowth influencers peddling a theory of "radical abundance."

So why? Well, finance capital wins no matter what plays out.

If demand-focused rhetoric succeeds in keeping the working class attempting to adjust its demand in service of maintaining capital relations (as Marx pointed out Malthus's lampoon was intended to do), then it can't (or won't) organize some kind of power to leverage against the energy production sector or the capitalist ruling class itself. Instead, energy companies get to operate intermittent renewables they own, providing the image of a "green" future with a natural gas baseload they also own.

Or if idealist true believer degrowth environmentalists and their politicians were actually to achieve green austerity and it starves or freezes enough people to actually "degrow" energy demand, we must remember that these massive companies own the "green" infrastructure. They will have cornered a growing market, ironically enough.

Regardless, the current situation where an owning class holds control of everything is perpetuated. That's all that matters here. Everything is in service of that. If that was possible to maintain with intermittent electricity and WITHOUT any kind of austerity, so people would be comfortable and not complain, we'd already be doing that.

It's not, though; there are both ideological and mechanical contradictions that are impossible to ignore. So the dichotomy they give us is "sacrifice the environment or sacrifice some people." They don't word it that way, but that's the choice. Which one people choose isn't something the imperial capitalist ruling class actually gives a shit about. They want different groups of people to make different choices and argue with each other. They actually don't want a definitive outcome.

Even they go back and forth, claiming they need to grow in one sector, degrow in another – then switch them. They have to! Cyclical crises will occur due to the inherent contradictions in capitalism. The falling rate of profit will introduce decay and hollow out industries while others surge into existence.

Either the capitalist ruling class really believes that they are the superior humans who deserve to exclusively consume the limited resources on this planet, to the point they will kill us (whether by policy or other means)... or they're cynically advocating for debunked crap that *would* kill us to keep us away from their wealth and position as the ruling class.

Either way, they're full of shit.

CONCLUSION

We're not supposed to be interested in contradictions. We're supposed to be interested in words like *"Wetiko,"* an Algonquin word for a mind virus driven by greed, excess, and selfish consumption that our society is supposedly sick with.

We're supposed to be stuck on what our neighbor is consuming rather than where the capitalist ruling class is taking us – or even that there *is* a capitalist ruling class that unaccountably makes all the decisions about society because they literally own everything.

And if that's getting us down, we're just supposed to want to destroy everything – and everyone! After all, what's a few lives when there's a surplus population? Less is supposed to seem like more. Less buildings, less pollution, less people. Burn it all down!

And maybe a total psycho follows through and bombs a building or something... not the typical route, but it does happen. More likely, they watch the ending of Fight Club and ugly cry. They watch the Ted K movie and think he was some misunderstood genius. They watch Contrapoints and spiral.

They're supposed to look out at the unwashed masses and think, "there's too many people making too many problems. And not much love to go around."

But I'm here to say this doesn't have to be a land of confusion.

We can't destroy everything – or even a lot of what currently exists. That is an insane thought. There's no way to start over, as many seem to want. History has happened. It's made its mark. America exists, and even if it didn't, tomorrow, its effects would remain. Whatever follows will retain some kind of American characteristics, even if it goes in the opposite direction – this is the dialectic of history.

We don't need fewer people. We need a different arrangement of power. We need a system of society that resolves the contradictions Marx and Engels identified in the 1800s while other idiots were telling everyone there were too many people to mask those contradictions in one way or another.

Clearly, that hasn't stopped. They keep trying to refine their rhetoric, so we don't notice they're telling us there are too many people on earth. But if the infrastructure that people use to live is reduced, then fewer people can live.

And some people truly believe this is the only path forward and outwardly say so, like Robert Jensen calling to halve the human population in his "Restless and Relentless Minds" article that Peter Buffett shared.

Some people cynically put it forward. Marx asserted that Malthus himself was possibly just working to propagandize against the French Revolution to justify the position of the newly-outdated feudal aristocracy. Many people assert that Margaret Sanger cynically put forward her racist, eugenic worldview and worked with KKK members (even putting one on her board) to "normalize birth control." And just to be clear, I personally consider birth control being acceptable to be a good outcome. But I can't say what was going on in her head. That bitch said some really fucked up shit, too. Fuck her.

Billionaire capitalists have for many, many years claimed to care about the "unsustainable population" for one reason or another. The Buffett Father and Son have talked about population a lot, Peter as the head of NoVo, putting money into various weirdos' pockets, Warren as a member of The Good Club, which a 2009 Guardian article called "the first time a group of donors of this level of wealth has met like that behind closed doors in what is, in essence, a billionaire's club."

Yeah it's the first time capitalists have met. I'm sure.

Oh wait, it's not, because that's exactly what The Club of Rome was, an organization that was founded in 1968 to advocate for a regulatory body made up of the world's biggest capitalists, presumably to give them more control to follow whatever plans they have.

Or, more bluntly, the World Planning Congress in Amsterdam in 1931, where the world's biggest capitalists met to propose various schemes to create a Planned Capitalism, ultimately impossible while retaining capitalism's defining contradiction of socialized production and privatized appropriation. If the point is to retain the ownership of capital that creates class, rational planning is impossible; resources get moved according to costs and profits, not quantities needed.

To truly plan global, imperial-stage capitalism, the anarchy of production has to be beaten. Outcomes must be enforced because otherwise, the falling rate of profit makes sustained planning impossible. What do people call that?

Fascism. Like, actually. No, not like, Umberto Eco listicle aesthetic bullshit. I mean real-ass Fascism. The thing that people like Rajani Palme Dutt, whom I've cited pretty extensively, and others have scientifically researched, excavating a defining character.

Fascism is the only coherent result of the fact that the form of private ownership of the means of production we call Capitalism can no longer develop. To continue to exist uninterrupted, the system must generate violent crises, stagnation, and decay. Fascism is an advanced stage of capitalism in crisis. As this world crisis grows, the number of unemployed, desolate people goes up while it becomes more and more necessary to lower the costs of production.

As I said, capitalist exploitation can no longer be coerced; it can only be enforced.

Degrowth's reactionary lineage through insistence on the preservation of Capitalism's ownership dynamics (whether fully intentional or conveniently ignored, promoted by the capitalist ruling class), back through idealist "Small is Beautiful" environmentalism and eugenics, the determination and designation of what parts of the population we cull through various pseudosciences, all the way back to Robert Thomas Malthus and his defense of stuffed-shirt aristocrats who spent all day fucking their cousins' brains out...

This reactionary lineage is telling us that the problems of the world are the fault of our neighbors. They're telling us that we are better than our neighbors but worse than the rulers – and whatever they decide is always the right thing to do. They got there because of God or the Invisible Hand; they're the best and smartest, and their asset holdings and bank accounts are the evidence we're supposed to accept.

Yes, capitalism requires growth – but not all of the time. The falling rate of profit is also going to require the limitation of production at other times. Thus, Capitalism also requires degrowth. Imperial-stage capitalism just cycles back and forth between crises and decay – one always being the way out of the other.

We have lived in the era of Imperial Capitalism for at least a century, and it has cycled in exactly this way. The cycle is decaying more and more, and what we see now is the result of decades of crisis, decay, and war, the ultimate resolution of both.

As long as we do not progress to a higher stage of production, one which resolves capitalism's fatal flaws that create class and nullify the human talent to plan with our brains, this is how it's going to be.

Humanity isn't the problem here. Limited resources are not the problem. Carbon production *isn't even the problem*. If capitalists could handle moving to a higher form of energy production like nuclear, which produces literally zero emissions... we'd already have done it.

The problem is that this system *can not accommodate doing things purely for the sake of making the world better*. Either it's got to be for profit or for the maintenance of class too.

And in that way, humanity is also the solution. Most of us know this is bullshit. We may not all understand why and a lot of us are misled because we've only ever heard lies. But we're here, right?

That's the best part about humanity: we're here. There are a whole lot of us. And most of us are not in the ruling class. Like, so many that the tiny numbers of people in society's ruling classes have spent centuries promoting various personalities, thinkers, and rhetoric saying, "there are too many people."

Humanity is the solution. We can all live better lives, we can all live in abundance. None of this bullshit is real! It's been disproven and debunked over and over and over and over.

Of course, the ruling class thinks there are too many people. We outnumber them by billions.

REFERENCE

[1] Meadows, Donna et al, "The Limits to Growth," Club of Rome (1972) http://www.donellameadows.org/wp-content/userfiles/Limits-to-Growth-digital-scan-version.pdf

[2] *5 facts about food waste and hunger: World Food Programme.* UN World Food Programme. (n.d.). Retrieved November 29, 2022, from https://www.wfp.org/stories/5-facts-about-food-waste-and-hunger

[3] Harris, P. (2009, May 30). *They're called the Good Club - and they want to save the world.* The Guardian. Retrieved November 29, 2022, from https://www.theguardian.com/world/2009/may/31/new-york-billionaire-philanthropists

[4] Marx, K. & Engels, F. (1846). *The German Ideology.*

[5] Crawford, et al, "Loci associated with skin pigmentation identified in African populations," *Science* (2017) https://www.ncbi.nlm.nih.gov/pmc/articles/PMC5759959/

[6] "Vacher De Lapouge and the Rise of Nazi Science," Jennifer Michael Hecht http://www.jstor.org/stable/3654029

[7] Robert Golub and Joe Townsend, "Malthus, Multinationals and the Club of Rome. Social Studies of Science" Social Studies of Science (1977) https://www.jstor.org/stable/284875

[8] Julian Simon, "Why Are Forecasts So Often Wrong?" from *The Ultimate Resource 2* (1981, updated 1996) https://

www.google.com/books/edition/The_Ultimate_Resource_2/0dLgDwAAQBAJ?hl=en&gbpv=1&dq=The+Club+said+that+the+conclusions+of+that+first+report+are+not+correct+and+that+they+purposely+misled+the+public+in+order+to+%E2%80%9Cawaken%E2%80%9D+public+concern&pg=PA49&printsec=frontcover

[9] Paul Dragos Aligica, "Julian Simon and the 'Limits to Growth' Neo-Malthusianism," *The Electronic Journal of Sustainable Development* (2009) 1(3) https://mercatus.org/uploadedFiles/Mercatus/Publications/JULIAN_AND_THE_LIMITS_TO_GROWTH_NEO-MALTHUSIANISM.pdf

[10] https://shrinkthatfootprint.com/average-household-electricity-consumption/

[11] https://rpsc.energy.gov/energy-data-facts

[12] Jason Hickel, Martin Kirk, "Don't Be Scared About The End Of Capitalism—Be Excited To Build What Comes Next," Fast Company (2017) https://www.fastcompany.com/40454254/dont-be-scared-about-the-end-of-capitalism-be-excited-to-build-what-comes-next

[13] Robert Jenson, "Restless and Relentless Minds: Thinking as a Species Out of Context" Resilience, a Post Carbon Society Project (September 4, 2021) https://www.commondreams.org/views/2021/09/04/restless-and-relentless-minds-thinking-species-out-context

ABOUT THE AUTHOR

Peter Coffin

Peter Coffin is a video essayist (Very Important Documentaries) with over a quarter-million YouTube subscribers, podcaster (PACD), and author (Custom Reality and You). Relatable humor and a commitment to everyday people keeps their perspective fresh, fun, and most importantly sharp.

@petercoffin

www.ingramcontent.com/pod-product-compliance
Lightning Source LLC
Chambersburg PA
CBHW070447220526
45466CB00004B/1778